Representations of the post/human

Monsters, aliens and others in popular culture

Elaine L. Graham

Manchester University Press

Published by Manchester University Press
Oxford Road, Manchester M13 9NR, UK
www.manchesteruniversitypress.co.uk

British Library Cataloguing-in-Publication Data
A catalogue record for this book is available from the British Library

ISBN   0 7190 5441 9 hardback
        0 7190 5442 7 paperback

First published 2002

10 09 08 07 06 05 04 03 02   10 9 8 7 6 5 4 3 2 1

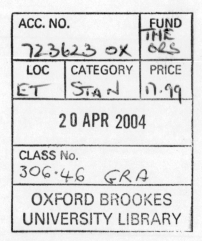
Typeset
by Graphicraft Limited, Hong Kong
Printed in Great Britain
by Biddles Ltd, Guildford and King's Lynn

CONTENTS

Contents

LIST OF FIGURES

Published in collaboration with the Centre for Religion, Culture and Gender at Manchester University, this innovative series addresses questions relating to the study of religion in Western culture. To what extent has this study been engendered? How do the disciplines within religious studies and theology serve a disciplinary function within society? What is the relationship between power, gender and knowledge in Western culture? What are the religious implications and tensions of the emphasis on 'the body' in postmodern and feminist theory? By making these questions central to the series a new direction is forged, which destabilizes traditional disciplinary boundaries while retaining academic rigour.

In this volume, Elaine Graham traces the configurations of religion, culture and gender in the conception of the posthuman. The understandings of humanism and of the human which grounded modernity have been destabilized. The posthuman is ourselves: 'we are all cyborgs now'. Rapid changes in technology, whether genetic or digital, signal the need to ponder the implications for culture and gender in more sophisticated ways than does the debate about 'playing God', a debate that uses overtly religious language but often does not recognize the underlying religious motifs for the conceptualization of the post/human. Elaine Graham investigates the cultural symbols of modernity, from Frankenstein to Star Trek, and from popular science to state-of-the-art technology, to develop a thoughtful profile of possibilities for the posthuman, ethical, cultured, and gendered.

On cathedrals, canals and computers

At eleven o'clock on the morning of 21 June 1948, in a workshop in a tiny side street on the campus of the University of Manchester, a team of mathematicians and engineers conducted the world's first successful electronic stored-program computing sequence on a machine they named Baby. Since then, the size of computing machines has reduced from a mainframe structure, the size of a lecture theatre, into desktop, even hand-held personal machines. Their speed and power has increased millionfold. Computer-mediated communications have transformed the abilities of users to store and process information, but more widely, have also changed leisure habits, the jobs many people do, and the types of machines that fill offices, shops and homes. Together with the identification, half a decade later, of the structure of DNA, the birth of Baby has triggered a technological and cultural revolution.

The city of Manchester itself has always been living witness to the close interaction between technological and cultural change. A century before Baby, Manchester had an international reputation as 'Cottonopolis', the first great city of the modern era. Its notoriety attested to the manufactories' transformation of economic activity, social relations, religious practice and political culture (Engels 1958). Now, half a century after Baby, in the opening years of the twenty-first century, the local economy struggles to reinstate its former glories within a very different, global, economic order in which – partly as a result of the digital revolution wrought by computer technologies – the chief source of economic power and wealth is information, encoded and transmitted instantly and cleanly at a keystroke.

Local contexts tell a story, therefore, and it has been partly as a result of reflecting on the changing fortunes of the city where I live and work that this book has taken shape. In the Victorian age, the material circumstances of economic transformation were symbolized in the towering gothic buildings erected as secular cathedrals to civic pride. 'Cottonopolis' encapsulated not only economic pre-eminence but the emergence of a new social order, the birth of class and the crowning of the modern era. Today, at the beginning of a new millennium, we search hard for similar symbols to express our aspirations. In 2002, Manchester will host a major athletics competition, the Commonwealth Games, an event upon which many are depending for a revitalization of the city's fortunes. The great monuments to Manchester's vanished ascendancy still stand, but alongside them have arisen new testimonies, if not to civic pride, then to what is expansively called 'urban regeneration'. Times are changing, and economic and social reconstruction now rests on post-industrial activities such as sport, heritage and leisure. The familiar cathedrals, canals and factories have been joined by shopping centres, café-bars and concert halls.

Visible symbols, such as architecture, iconography or the corporate logo can speak volumes about the deeper values and aspirations of a particular institution. So it is that I have chosen to concentrate on the outward signs of deeper technological, economic and cultural change; how the pressures and transitions of living in a digital and biotechnological age are being expressed in the exemplary images, metaphors and icons of Western culture. However, it is also necessary to reach behind superficial appearances and interrogate the implicit metaphysical and theological values embedded in these representations: what we worship, where we place our ultimate trust, what it means to be human.

In undertaking this study, I hope to emulate the achievements of my two predecessors in the Samuel Ferguson Professorship of Social and Pastoral Theology, Ronald Preston and the late Tony Dyson. Both of them held the conviction that the credibility and effectiveness of theological scholarship could only be conducted in critical dialogue with the realities of contemporary social, economic and scientific developments. Ronald's work on the interaction between theology and economics established him as a leading figure in twentieth-century Christian social thought; and in his engagement with the emergent biotechnological and genetic disciplines, Tony combined respect for the integrity of the human and natural sciences with rigorous theological analysis. I share their commitment to discerning the signs of the times amid current scientific and cultural trends; and my hope is that I can continue this tradition of constructive theological conversation with the disciplines and institutions of contemporary society.

I should like to express my gratitude to all those who have been instrumental in the creation of this work. The University of Manchester Research Support Fund enabled me to take a period of study leave in 1998–99. The collegial environment of the Department of Religions and Theology at Manchester is second to none, and I am indebted to my colleagues for their encouragement in the execution of this research. In particular, I would like to mention Philip Alexander, John Atherton, George Brooke, Adrian Curtis, Bernard Jackson, Todd Klutz and Graham Ward, all of whom have been valuable and stimulating conversation partners at various stages in the process.

My thanks go also to three former students, Malcolm Brown, Frances Ward and Tracey Low, whose intellectual concerns have helped to clarify my own thinking.

Further afield, I am grateful for support and assistance of various kinds from John Armes, Alison Adam and Eileen Green, Tina Beattie, Heather Walton and Sharon Welch, and to the Science and Religion Forum, Queen's College, Birmingham, Trinity and St John's Colleges, Auckland, New Zealand and the Manchester Theological Society to whom I had the opportunity to present versions of work in progress. Special thanks go to Colin Avery and Peter Gomola for their contributions to Chapter six, and to Grace Jantzen, series editor and Publications Director for the Centre for Religion, Culture and Gender for her help and guidance. Finally, I should like to dedicate this book to the memory of the late Carol Smith, whose encouragement, practical and intellectual, was crucial in getting this project off the ground.

Mapping the post/human

Before the next [twenty-first] century is over, human beings will no longer be the most intelligent or capable type of entity on the planet. Actually, let me take that back. The truth of that last statement depends on how we define human . . . The primary political and philosophical issue of the next century will be the definition of *who we are*. (Kurzweil, 1999a: 2, my emphasis)

On being human

This is a book about what it means to be human. Or, more precisely, it is an examination of the impact of twenty-first-century technologies – digital, cybernetic and biomedical – upon our very understanding of what it means to be human. I will argue that some of the most definitive and authoritative representations of human identity in a digital and biotechnological age are to be found within two key discourses: Western technoscience (such as the Human Genome Project) and popular culture (such as science fiction). To interrogate such representations is to draw out the implicit desires, anxieties and interests that are fuelling humanity's continuing relationship with its tools and technologies. The 'worlds' engendered by the creative imagination – in myth, language and religion – are just as revealing, in their own way, of the ethical and political dimensions of the digital and biotechnological age as are the material artefacts of humanity's technological endeavours.

There has always been a close relationship between technological innovation and social change. The great spinning machines and the manufactories engendered new economic relations during the industrial revolution, and effected new patterns of work, social class and urban life. Yet new technologies have done more than simply introduce new patterns of work, leisure and social interaction; they have called into question the immutability of boundaries between humans, animals and machines, artificial and natural, 'born' and 'made'.[1] In challenging the fixity of 'human nature' in this way, the digital and biotechnological age engenders renewed scrutiny of the basic assumptions on which matters such as personal identity, the constitution of community, the grounds of human uniqueness and the relationship between body and mind are founded.

New technologies have complicated the question of what it means to be human in a number of ways:

TECHNOLOGIZATION OF NATURE

In the splice-and-dice biotechnology laboratories, genetic material no longer has the fixed, immutable status it used to occupy in the theory books of destiny. (Ross, 1996b: 18)

An examination of the extent to which molecular biology and genetics since 1953 (the date of the identification of DNA) have transformed scientists' ability to intervene in processes of conception and gestation through new reproductive technologies and in genetic modification of plants, animals and even humans, may tempt one to think of the past fifty years as an unprecedented period. However, while we may be more aware than ever that what we call 'nature' is open to manipulation by new reproductive technologies, genetic modification and other biotechnologies, greatly enhancing the sophistication of scientific and medical manipulation of nature, it is important also to realize that what we call 'nature' has always been subject to human intervention. Humanity has long intervened in the development of other living organisms for its own purposes, practising rudimentary forms of genetic modification through the selective breeding of plants and animals (Reiss and Straughan, 1996: 1–7). While living species have evolved gradually over millions of years, therefore, and human intervention in what we call 'nature' is an everyday part of what we call 'culture', contemporary biotechnologies nevertheless offer unprecedented opportunities greatly to accelerate the pace of change, even to effect the emergence of new species, as a result of direct and deliberate genetic manipulation (Peters, 1997: xiii–xvii). Growing sophistication in biomedical techniques such as assisted reproduction and rudimentary cloning are ready indications of how scientists are now routinely able to intervene in so-called 'natural' processes with the aid of technologies. Genetic modification of plants and non-human animals also suggests that biology becomes 'an artifact of technology rather than its limit' (Cole-Turner, 1993: 9).

---

[1] Clearly, the consequences of such advances, especially in genetics, go beyond clinical practice, social policy and technoscientific research. They have also generated new debates in bioethics among moral philosophers and theologians, as the ethical implications of new scientific methods become clear. Many of these will already be familiar, such as the use of foetal tissue for experimentation, the moral status of a foetus conceived via in vitro fertilization, the legal implications of surrogacy, and the consequences for our rights to privacy, autonomy and informed consent, and so on (Nussbaum and Sunstein, 1998; Cole-Turner, 1993; Glover, 1984; Harris, 1998).

The epitome of what might be termed the technologization of nature, is the digital artificial life system, known as 'A-Life'. Thomas Ray's 'Tierra' A-Life programme (Hayles, 1999: 223–31), consists of computer simulations of elementary organisms to see how they grow, reproduce and evolve through 'genetic mutations' that have been written into the software program. The key narrative underpinning the design of Tierra expresses a commonplace scientific metaphor, widespread across much of contemporary technoscience, of the essence of life as a flow of information. Tierra's design effectively transposes the dynamics of natural selection into the design of a synthesized genetic software program, and represents the translation of patterns of biological carbon-based life into models of silicon-based information technologies.

BLURRING OF SPECIES BOUNDARIES

New reproductive technologies, cloning and genetic modification also promise to engender a future in which the boundaries between humanity, technology and nature will be ever more malleable. Recent debates about genetically modified crops and transgenic treatments of animals reflect biotechnological developments that enable genetic material to be transferred from one species to another. There has also been discussion of using non-human animal organs in human transplants. Some Western governments have granted legislation permitting limited experimentation for therapeutic purposes into the use of human embryonic tissue cloned from adult cells. It is envisaged that this will assist in healing tissue following organ transplants or skin grafts by minimizing the risk of rejection (National Bioethics Advisory Commission, 1998; Nuffield Council on Bioethics, 1999).

In the area of artificial intelligence, the phenomenology of human–machine interaction maps the implications of computers becoming more interactive and sophisticated (Turkle, 1995; Foerst, 1996; Adam, 1998). If this heralds a new age of smart machines that can simulate, or even surpass, human intelligence (Kurzweil, 1999a), then what are the effects of taken-for-granted understandings of human uniqueness when machines seemingly appear to be replicating human functions?

TECHNOLOGIZATION OF HUMAN BODIES AND MINDS

Once the province of military experiment and science fiction, 'cyborgs' (fusions of the organic and cybernetic) are also moving from fantasy into reality (Gray, 1995a). Miniaturization of components such as microchips means that it is possible for technologies now to be placed into or grafted on to the human body, enabling a greater degree of functional integration between human and machine (Warwick, 1998). The technologization of the human body is already taking place in many medical contexts: to power prosthetic limbs scientists are experimenting with electronic transmitter implants that will stimulate nerve endings. Even things like heart pacemakers, artificial retinas and cochlear implants are becoming commonplace, at least in the West (Evans, 1998).

CREATION OF NEW PERSONAL AND SOCIAL WORLDS

Computer-assisted technologies are also transforming many aspects of contemporary Western life. The typical working office could not function without recourse to information technologies such as personal computers, mobile phones, modems and faxes. Facilities such as cheap internet access and personalized digital radio/TV schedules, assembled by computer program to suit individual preferences, are transforming assumptions about patterns of leisure as well as work. Information is arguably the most prized commodity in the global economy of the twenty-first century, increasingly the major source of wealth and undoubtedly a determinative mode of power. Social change is no longer driven by the dynamics of material production and consumption, but by the circulation and commodification of *data*.

Just as the boundaries between humans, animals and machines are eroded, so too are distinctions between the virtual and the real. New digital technologies have reconfigured taken-for-granted patterns of physical space, communication and intimacy. Digital technologies will soon be of such sophistication as to be capable of synthesizing an entire virtual environment, creating cyberspace into which a person can be sensorially if not physically assimilated. The distances of time, space and place are shrunk, as computers enable their users to call up resources on the other side of the world (Baudrillard, 1983; Virilio, 1989). The introduction of information technology is, arguably, on a par with the invention of the printing press in its capacity to revolutionize patterns of communication. As many cultures have evolved historically from oral to written, so too print-based media such as books may be displaced by new forms of communication, with corresponding adaptations in styles of reading, textual analysis and attention spans.

TOOLS, BODIES AND ENVIRONMENTS

Whereas it may once have been appropriate to think of tools such as the knife, the hammer and the water-pot as simple instruments of extension and containment, yet remaining separate from their users, in the highly technologized societies of the twenty-first century the very boundaries between the human and the machinic is redrawn. There are two dimensions to this. Firstly, prostheses, implants and synthetic drugs are internalized by the organic body. Technologies are not so much an extension or appendage to the human body, but are *incorporated*, assimilated into its very structures. The contours of human bodies are redrawn: they no longer end at the skin. 'The skin has been a boundary for the soul, for the self, and simultaneously a beginning to the world. Once technology stretches and pierces the skin, the skin as a barrier is erased' (Stelarc, in Atzori and Woolford, 1995).

A second characteristic of the digital and biotechnological age is its *immersive* nature, as new habitats for the containment and circulation of communication are created. Digitally generated virtual worlds offer new forms of 'post-bodied' activity, and the unitary face-to-face self is superseded by the multiple self – the simulated, fictive identity of the electronic chat-room or the multi-user domain, the avatar or synthesized self of a digitally synthesized interactive environment. Virtual reality allows the user to project a digitally generated self into cyberspace, synthesizing new spatial and temporal contexts within which alternative subjectivities are constructed. In this context, digital technologies are being used not so much as *tools* – extensions of the body – as total *environments* (Poster, 1996: 188–9). Such digitally generated virtual domains 'cannot be viewed as instruments in the service of pre-given bodies and communities, rather they are themselves contexts which bring about new corporealities and new politics corresponding to space-worlds and time-worlds that have never before existed in human history' (Holmes, 1997a: 3).

New relationships between 'the human, the natural, or the constructed' (Haraway, 1991c: 21), therefore reveal the very categories 'humanity', 'nature' and 'culture' as themselves highly malleable. Technologies call into question the ontological purity according to which Western society has defined what is normatively human. For example, in *Blade Runner* (1982), Ridley Scott's film adaptation of Philip K. Dick's novel *Do Androids Dream of Electric Sheep?* (1968), a species of androids, called 'replicants', has escaped from off-world space settlements and come back to earth. Replicants are machines that are superficially indistinguishable from humans. At the same time, however, humans have come more and more to resemble machines in their high-tech, alienated, urban wasteland surroundings. The fragile and indeterminate nature of the very boundary between humans and artefacts, sentience and inertness, authentic and artificial, constitutes the heart of the novel and its film adaptation (Lyon, 1999: 1–6).

As the opening quotation from Ray Kurzweil indicates, therefore, it is not only a question of coming to terms with the economic and cultural impact of new technologies, but of engaging with their capacity to stir up questions of *ontology* (Holmes, 1997a: 9). To place the contours of human nature under such pressure, however, is also to invite a range of reactions, positive and negative. New technologies are often perceived as threatening bodily integrity, undermining feelings of uniqueness, evoking feelings of growing dependency and encroachments on privacy. It is easy to portray such trends as the 'homogenization of the human by the technological' (Brasher, 1996: 815), an erosion of the taken-for-grantedness of human physical and psychological integrity by invasive, deterministic technologies. And yet there is a parallel strand that sees technologies as enhancing lives, relieving suffering and ushering in the unlimited flourishing of human potential; of the digital and biotechnological age as facilitating a period of human empowerment and evolution – even divinization.

5

Enslavement or liberation?

Although reactions to technologies are frequently polarized into either the technophobic or technophilic,[2] these broad alternatives conceal a diversity of understandings of the nature of what is meant by technology and how its social and ontological impact may be assessed. The range of responses to different kinds of technologies, practical and philosophical – as well as a recognition of the diverse nature of technologies themselves – may therefore more usefully be ranged along a continuum.

'DISENCHANTMENT'

For some, the impact of new technologies signifies the dissolution of the distinctively organic human into a variety of engineered, hybrid, modified or virtual conditions. Far from assisting human development, technology will bring about alienation and dehumanization, the erosion of the spiritual essence of humanity. This is a perspective frequently associated with the sociologist and theologian Jacques Ellul (Ellul, 1965; Feenberg, 1991: 4–5). Ellul drew upon Max Weber's association of modernity with the intensification of reason as the predominant mode of human social action and organization. With that goes the erosion of the affective and traditional realms of social action in favour of the rule of rational–technical instrumentality. Ellul regarded technology as exerting an almost demonic tyranny over humanity and nature, its inherent rationality doing violence to the sacredness of existence. For Ellul, however, 'technology' is not a matter of certain machines in particular, or specific processes of manufacturing, such as mechanization or automation, but rather a matter of 'technique', which Ellul defined as 'the totality of methods rationally arrived at and having absolute efficiency (for a given stage of development) in every field of human activity' (1965: xxv). Ellul thus castigates a world in which the rationality of technique – what might be termed the cult of efficiency – reigns supreme: 'Man [sic] is not adapted to a world of steel; technique adapts him to it . . . when technique enters into every area of life, including the human, it ceases to be external to man and becomes his very substance . . . it progressively absorbs him' (Ellul, 1965: 6).

While his work is of a very different calibre to that of Ellul, Martin Heidegger's writing has dwelt on similar themes in the philosophy of technology. Heidegger feared that technologies would bring about what he called the 'darkening of the world' (1993a: 331). Heidegger claimed that technology was not to be understood or approached through its design or function, or even in terms of productive processes, but rather in its potential to 'enframe' (*Gestell*) or yield the inner essence of being (1993a: 324). More than mere tool, technology was the very

---

[2] As in the representations of 'Puppet' and 'Prometheus' (Peters, 1997) or 'enslavement' and 'liberation' (Cooper, 1995).

vehicle by which being is brought forth (*alétheia*) into tangible form (1993a: 319). Eventually, however, the logic of technological framing would so dominate that it would obscure alternative revealings of the nature of being, and technology would transform everything, including human labour, into 'standing reserves' (*Bestand*) or raw materials awaiting appropriation. 'Everywhere everything is ordered to stand by, to be immediately on hand, indeed to stand there just so that it may be on call for a further ordering' (Heidegger, 1993a: 322). Creative activity would become so contaminated by the forces of reification that the inner truth of artefacts was obliterated rather than realized (Heim, 1993b: 61).

As Heidegger once remarked, then, 'the essence of technology is by no means anything technological' (1993a: 311), a statement which identifies in technology not only an empirical reality, but a force that may more appropriately be termed ontological. It hints that technologies are more than empirical objects or tools, but possess the capacity to constitute human existence, a philosophical perspective that Andrew Feenberg characterizes as a 'substantive' model of technology (1991: 5–6). However, it is important to note at this stage that technologies may be constitutive or substantive, even determinant, without being deterministic. This point is what distinguishes Heidegger's views from those of Ellul, and begins to identify how the category of 'technophobia' may be a highly misleading category by which to make sense of certain representations of the digital and biotechnological age. Ellul sees no escape from the inexorable 'iron cage' of rationality, but reduces 'technology' to an abstract, monolithic, culturally universal force, and effectively displaces the materiality of technologies into the function of an abstract reason, as well as being simplistic in its demonization of technological advances. By contrast, Heidegger's later work does articulate possibilities for reconfiguring human relationships to technologies. The solution lies in building up what Heidegger refers to as a 'free relationship' to technologies, a relationship that will be free if it 'opens our human existence to the essence of technology' (1993a: 311). While insisting on the ontological nature of technologies – a power to build worlds and selves – Heidegger also argues that it may be possible to establish a relationship between humanity and technology that is not nihilistic. He thus conceives of technologies as an integral aspect of human becoming – for good or ill – but resists a crude determinism about their power to dehumanize and objectify.

This perspective is nevertheless helpful as an exposure of tendencies to relate to tools and artefacts that effectively reduce everything to a factor of human self-interest. However, it highlights the importance for any analysis of the digital and biotechnological age of the need to differentiate between technologies and to be mindful of the role of human agency, the techniques that are deployed, the quality of the artefacts and worlds created, and approaches to the management of the natural and artificial environments, as necessary properties of human engagement with technologies.

'TOTALITARIANISM'

This perspective also expresses misgivings about the potential of technologized society to suppress democracy, erode individuality, invade the private realm and encroach upon face-to-face contact, embodied interaction, indigenous cultures, traditional forms of community, print-based media or civic values in the face of voracious digital and virtual communications. As Albert Borgmann argues, the subtleties of interpersonal relationships are attenuated by the 'hyper-texted' society (Borgmann, 1984: 196–210).

Similarly, Jürgen Habermas' work has been adapted into the philosophy of technology as representing an analysis in which technological development is equated with the crushing not only of the human spirit but with all aspects of the non-rational, the spontaneous, the human(e). Technologies – again, represented by the intensification of instrumental reason – diminish the arena over which people can exercise self-determination. As the public sphere becomes more aligned to rational–technical institutions, so the life-world of community, democracy and ethics is invaded by a culture of domination (Habermas, 1970; Feenberg, 1991: 5–6).

'TECHNOCRATIC'

Exemplified by the futuristic optimism of popular science writers such as Michio Kaku and Ray Kurzweil, and the advocate of virtual reality Howard Rheingold (all featured in Chapter seven), this view also represents a recurrent thread throughout much twentieth-century Anglo-American science fiction (Chapters two and six). Advances in technology are regarded as politically neutral; and specific technologies are merely subservient to values and objectives established elsewhere. Technological developments are also entirely universal and transferable, in that particular cultural values and economic interests are in no way embedded in the design or use of technologies.

In contrast to the 'substantive' or constitutive model, such an understanding of the relationship between humanity and its tools might be described as 'instrumental' (Feenberg, 1991: 5). Technologies are tools, existing merely to carry out the will of their maker(s). When predicting the shape of a future world order in the digital and biotechnological age, however, technocratic futurism fails to address questions of access to technoscientific resources and opportunities, a vision of the future in which the 'gadgets may be different, but social relations remain static' (Robins and Webster, 1988: 49).

TRANSHUMANISM

Technocratic optimism is carried further within analyses that look to the technological developments as promising the future *evolution of Homo sapiens*. With the aid of technological enhancements, human beings will attain immortality and omnipotence (Regis, 1990; More, 1998). The term 'transhuman' is a conflation

8

of transitional human, or one augmented and modified on the way to being *posthuman*, the fully technologized successor species to organic *Homo sapiens*.

Transhumanism celebrates technology as the manifestation of human liberation from bondage to nature, finitude, and the vagaries of disease, decay and death (Cooper, 1995: 12). Note, however, how a particular response to technological innovation articulates a very specific model of what it means to be human: technologies not only represent protection from that which threatens physical survival, but are a means of 'transcending' those physical limitations altogether. Whether the body is augmented, rebuilt or obselete (as in the work of Hans Moravec (1988; 1998) who predicts that human intelligence will eventually be downloaded into computer hardware), the essential, rational self endures unimpeded. Stephen Clark (1995) argues that much science fiction is imbued with transhumanist sentiment, driven by a desire for the subjugation, even the effacement, of vulnerability, contingency and specificity. Yet transhumanism betrays a doctrine of humanity informed fundamentally by a distrust of the body, death and finitude, issues which, as I shall indicate, have ethical, political and theological implications.

RE-ENCHANTMENT

This final perspective mirrors my first category, although rather than technologies being portrayed as depriving the world of spirit, they are represented as the vehicles of ascent to a higher plane, the artifices of divinity, a harnessing of elemental powers. Proponents of what has been termed 'technochantment' (Bennett, 1997: 17) seize upon digital and virtual technologies as helping to reinject the spiritual, the religious and the 'transcendent' into a materialist world, while fully (and often uncritically) embracing technoscientific advances. The rhetoric of cyberspace as sacred space (Wertheim, 1999; Heim, 1993a, 1993b) and the 'technologization of the ineffable' (Lieb, 1998) seemingly resurrect a transhistorical expression of an inherent human desire for 'transcendence' (Noble, 1999). This represents not only the enhancement or evolution of *Homo sapiens*, but its very divinization.

I intend to discuss how all these various representations of the digital and biotechnological age are manifested in the Western cultural imagination as my argument unfolds. Whether the rhetoric be one of aggrandizement or obliteration in the face of technological proliferation, however, there is still an implicit assumption that such a process is inexorable and inevitable, spearheaded by an inexorable, monolithic force called 'technology' in the abstract. The logic of 'evolution', in which humanity is superseded by a successor species – machinic, superhuman or a combination of the two – regards the transition from *Homo sapiens* to postbiological *Homo cyberneticus* as a foregone conclusion; and the narrative of disenchantment fails to consider the political, economic and cultural choices embedded within the design, appropriation and distribution of new technologies.

9

However alienating, impersonal or soul-destroying they seem, therefore, technological processes always entail human agents.

Definitions of technologies as 'nonorganic crafts, tools, and machines created by humans' (Springer, 1996: 15), or as 'a rational, systematic, taught, learned, and replicable way of materially controlling the material world, or part of it' (Russ, 1995: 35), begin to offer alternatives to monolithic and abstract characterizations. They enable us to consider technologies as reflexive phenomena; that is, as culturally mediated systems, the products – albeit institutionalized and, frequently, used to inequitable ends – of human *agency*, but which in turn exercise a potentially constitutive power over human experience, for good or ill. Technologies are, as Bruno Latour might have it, always already full of human labour (Latour, 1993).

Full of human labour they may be, and profoundly practical instruments of extension, containment and manipulation too. But, like any cultural artefact, tools and technologies are also bearers of meaning. It is impossible to abstract technologies from their social and symbolic contexts. Technologies are not just the 'application' of pure science. Nor, for that matter, is science independent of culture. Pure science cannot be divorced from applied technologies, because social and political priorities help determine the conduct of research. Far from being an elevated and disinterested intellectual pursuit, scientific funding, research and application are intimately connected to the logic of transnational corporate capitalism (Haraway, 1997; Aronowitz et al., 1996).[3] The term 'technoscience' (Ross, 1991: 1–15; Haraway, 1997: 279–80), thus encapsulates both the cultural embeddedness of things and the material and practical nature of ideas. Its usage serves as a reminder that scientific practices and institutions do not simply observe, but actually participate in the construction of knowledge, a thesis I shall pursue in Chapter two when I pursue further the understanding of technoscience and popular culture as mutually reinforcing systems of *representation*.

Representing the post/human

Whether the biotechnological and digital age is conceived as establishing a 'satanic mill' of dehumanization or furthering the realization of 'celebrations of the technological sublime' (Penley and Ross, 1991a: xii), it is clear that radical reappraisals of the future of the human species are being advanced. Echoing Lyotard's celebrated work *The Postmodern Condition* (1984), Halberstam and Livingston refer to the advent of the 'posthuman condition' (1995: 19), denoting a world in which humans are mixtures of machine and organism, where nature has been modified (enculturated) by technologies, which in turn have become assimilated

---

[3] The commercial interests apparently at work in the funding and patenting of the Human Genome Project is one clear example (see Chapter five).

into 'nature' as a functioning component of organic bodies. Similarly, Featherstone and Burrows speak of the complicity of new technologies not just in the 'making and remaking of bodies, but the making and remaking of worlds' and refer to the future of humanity as 'post-bodied and post-human' (1995: 2) (see also Hayles, 1999; More, 1998).

Such terminology raises a number of questions, however. I have already identified a spectrum of diverse responses to new technologies, ranging between the digital and biotechnological age as one of enslavement or liberation, or advanced technologies as threats to human integrity or means of facilitating its further evolution. This generates an imperative to interrogate more deeply the values and interests that underpin any representation of the 'posthuman condition'. What is at stake, supremely, in the debate about the implications of digital, genetic, cybernetic and biomedical technologies is precisely what (and who) will define authoritative notions of normative, exemplary, desirable humanity into the twenty-first century. There is also a question of how such visions will be enshrined in the design of technologies and built environments; how they will shape political and policy choices about scientific funding; and how they will inform scientific theories and metaphors. It is not so much about the end of the human or the advent of the superhuman, therefore, as whose visions will fuel the techno-scientific developments that contribute towards the realization of any such futures. For this reason, therefore, I have not adopted the convention of the 'posthuman' common among other commentators, but have chosen the term 'post/human' instead. By this I hope to suggest a questioning both of the inevitability of a successor species and of there being any consensus surrounding the effects of technologies on the future of humanity. The post/human is that which both confounds but also holds up to scrutiny the terms on which the quintessentially human will be conceived.

The promise of monsters

While such questioning of understandings of 'human nature' may appear disquieting, however, there are critical tools to aid interpretation. Western culture may be confronting a technologically mediated 'crisis' of human uniqueness, but a more satisfactory way of framing the situation might be in terms of the blurring of boundaries, a dissolution of the 'ontological hygiene' by which for the past three hundred years Western culture has drawn the fault-lines that separate humans, nature and machines. Definitive accounts of human nature may be better arrived at not through a description of essences, but via the delineation of boundaries. Evidence for this recurrent convention in constructing discourses about what it means to be human may be discerned by tracing a *genealogy* of boundary-creatures. Human encounters with rudimentary technologies and the implications of their own inventive powers has long been the stuff of myth and literature. There have been fables in the West of talking and animated statues since antiquity, and thriving traditions survive of animal and human automata in China, strongly evident by the fourth or third century BCE. Accounts of automata also emanated from Greek, Arab and Indian writers from the same period (Mazlish, 1993: 32–3). Dreams of creating forms of artificial human life by other than heterosexual reproduction extend from the ancient Greek myth of Prometheus, through the Jewish legend of the Golem of Prague, through to *Frankenstein*. More recently, fictional robots, androids and smart computers offer us intriguing glimpses of machines transforming themselves from tools into sentient beings, with attendant questions about 'their' status in relation to 'us'.

One of the ways in particular in which the boundaries between humans and almost-humans have been asserted is through the discourse of 'monstrosity'. Monsters serve both to mark the fault-lines but also, subversively, to signal the fragility of such boundaries. They are truly 'monstrous' – as in things shown and displayed – in their simultaneous demonstration and destabilization of the demarcations by which cultures have separated nature from artifice, human from non-human, normal from pathological. Teratology, the study of monsters, bears witness to this enduring tradition of enquiry into the genesis and significance of the aw(e)ful prospect of human integrity transgressed. Significantly, teratology also straddled what we would think of as the disciplinary lines of religion and science, being simultaneously a theological and early scientific form of discourse (see Chapter two).

At the boundaries of humanity, machines and nature, the impossibility of fixed definitions is shown forth in the proliferation of contemporary signs and wonders. In their capacity to show up the 'leakiness' of bodily boundaries (Shildrick, 1997) this emergent array of hybrid creatures are arguably 'monstrous' not so much in the horror they evoke but in their exposure of the redundancy and instability of the ontological hygiene of the humanist subject. These creatures – like early modern monsters, straddling the boundaries of fiction and scientific taxonomy – are 'processes without a stable object' (Braidotti, 1996a: 150). As the categories 'human', 'nature' and 'technology' are being radically reconfigured and intermingled, their miscegenation produces creatures of many kinds who continue the enduring debates about the limits and potential of human nature originating with the 'teratological' discourses of antiquity.

The work of Michel Foucault provides me with a useful heuristic framework for interrogating representations of the post/human. He destabilized essentialist and axiomatic accounts of human nature in favour of critical methods of *archaeology* and *genealogy* (Chapter two). Foucault was not concerned to state a fundamental truth about human nature so much as to elaborate the technologies of the self – such as practices of incarceration, confession and disciplines of the body – which he understood as the processes by which dominant models of human nature were thought and practised. He debunked the ontology of humanism. In his work – just like teratology – the pathological, the outcast, the abject and the almost-human consistently feature as indicators of the limits of the normatively human.

Michel de Certeau has argued that despite notional acknowledgement of the spaces for resistance to disciplinary regimes, Foucault's analysis effectively leaves little space for techniques of subversion and dissent (Buchanan, 2000: 100). Certeau insists upon the legitimacy of the procedures – 'many-sided, resilient, cunning and stubborn' (Certeau, 2000: 105) through which everyday practices could create a freedom, albeit circumscribed and ephemeral, from panoptic administration. However, it seems to me that Certeau has missed an opportunity to appraise the *vernacular* exercise of Foucauldian power. It is not that everyday, ordinary practices are 'liberative' and bureaucratic ones 'oppressive', for if as Foucault always claimed, power descended 'right down to the depth of society' (Foucault, 1977: 27), then it should be possible to imagine its effects circulating not just in authoritarian regimes of total surveillance, but also in the implicit values of commercialized popular culture, not least in the responses of its audience. My intention is thus to elaborate how popular and official representations are of a piece, and that both exemplify modes of enforcement *and* transgression.

Popular culture should therefore be considered as a significant site of the contemporary 'genealogy of subjectification' (N. Rose, 1996: 129). If the demarcation of modern selfhood can be undertaken within a diversity of institutions such as the clinic, the prison, the asylum and the confessional, then why not via the soap-opera, the virtual chat-line, the internet or the science-fiction fanzine?[4] A useful parallel to Foucault's genealogical critique, therefore, and one applicable to contemporary technoscience and popular culture, is what Marleen Barr calls 'fabulation' (see p. 58), or forms of popular culture that disrupt certainties or explode ideologies (1992). Fantastic, utopian and speculative forms of fiction – epitomized by science fiction – shock our assumptions and incite our critical faculties. As refractions of the same, as evidence for the ascribed and not essential nature of human nature, monsters, aliens and others provide clues for the moral economy or 'ontological hygiene' by which future categories of the human/posthuman/non-human might be decided.

If the boundaries between humans, animals and machines, or between organic and technological, are clearly under pressure in the digital and biotechnological age, then the relationship between another supposed binary pair, 'fact' and 'fiction', is also central to the argument of this book. The best contemporary illustration of this may be found in the use of the colourful epithet 'Frankenstein Food' in

---

[4] Surely something like cyberspace must be ripe for a Foucauldian analysis of how codes, protocols, commercial imperatives, modes of access and conventions of design all function to engender 'regimes' of virtuality in which self-identity is enacted, renegotiated and disciplined?

the media as a description for genetically modified crops (Monsanto Corporation, 1999; James et al., 1999; Wood, 1999). In this transgenic union of agribusiness and Gothic novel, we see how literature and myth still inform – for better and for worse – popular reactions to technoscientific innovation. *Frankenstein* continues to occupy a definitive role in the reception of biological and genetic innovation within popular media and public policy since its publication (Turney, 1998). This leads me to suggest that science and popular culture may both be regarded as *representations* of the world, in that both deploy images and rhetorical conventions which do not simply report reality, but construct, mediate and constitute human experience. Representational practices are part of the human activity of building material and symbolic worlds; of encompassing metaphysical and theological systems as well as cathedrals, canals and computers. My own analysis of discourses of the post/human takes its cue, therefore, from the need to examine scientific *and* literary, mythical and fictional representations as interconnected 'narratives' (Hayles, 1999: 22) that reflect and construct understandings of what it means to be post/human. Hence the title of Part I of the book, 'Science/fiction'.

Part II develops the key themes of monstrosity, genealogy and representation. Mary Shelley's *Frankenstein* is often identified as the first modern work of science fiction – the 'first great myth of the industrial age' (Aldiss, 1973: 23) – thus spawning an entire genre in which the literary imagination sought inspiration from science and technology. While the novel does articulate nascent anxieties about emergent scientific methods, it can only superficially be regarded as a chilling horror about a 'mad scientist'. More fruitfully, it offers reflections on the fundamentals of human identity, of humanity as created being and creative subject, life as organic or artifice, and the responsibility of human knowledge released from theological prohibition. In keeping with my themes of boundary-creatures as markers of discourses of what it means to be human, therefore, Chapter three argues that the question of the *attribution* of monstrosity is at the heart of *Frankenstein*'s critical reception. The representation of Victor Frankenstein's creature is multifaceted, as he is tested against conventional norms of exemplary humanity. Does his physical malformation disbar him from human society, or can he be redeemed by his ability to learn and acquire a cultured character? Do the monstrous circumstances of the creature's genesis condemn him to a non-human status? Victor's obsessiveness is framed by Mary Shelley as an inhuman(e) preoccupation with cheating death; is he, in his narcissistic obsession with controlling nature, the true monster of the tale?

Chapter four develops the theme of *genealogy*, and suggests that a single mythical figure inhabiting the boundaries of object and subject, creation and creature, can undergo many different cultural permutations. The golem has served as the refraction of many wider concerns about the nature of cosmogony and divine creativity, of Jewish identity, to issues of gender and technology. The golem's Biblical origins, and its evolution over time, offer a singular opportunity to trace a figure that has absorbed many changing cultural influences while remaining a potent medium through which to explore issues of demiurgical power and human responsibility.

Chapter five concentrates on the power of *representation* in shaping discourses about the post/human. It is helpful to see how notions of 'human' are themselves articulated and reproduced through technologies, literal and epistemological. Here, I develop the different aspects of the term 'representation' – as metaphor, as surrogate and as political mouthpiece. The political and commercial power of the Human Genome Project serves as a useful case study into the ideological nature of scientific discourse, alerting us to the currency of what Foucault termed 'bio-power' in articulating what it means to be human. If we are witnessing a new reductionism, in the form of the 'geneticization of human nature' as some critics suggest, then this represents a paradoxical attempt to reassert control over species identity at the very time when other trends suggest the dissolution of certainties regarding real or essential human nature, with significant political and ethical implications.

In Part III, 'Post/humanities', I consider further models of exemplary and normative humanity implicit in contemporary popular and technoscientific representation, thereby returning to philosophical debates about relations between human beings and their technologies. Chapter six discusses the *Star Trek* television series, often upheld as the exemplar of progressive technocratic optimism in its representation of the future. I examine the different approaches taken by two series, *Star Trek: The Next Generation*, and *Star Trek: Voyager*, towards the dilemmas engendered by forms of artificial life and cyborg technologies. While some characters, especially in the later *Voyager* series, explore greater latitude in technologized subjectivity, the dominant representation is still of a supreme rationalist subject uncontaminated by technologies.

Chapter seven examines other versions of technocratic futurism of a more expansive kind in which prosthetic, cybernetic and biomedical enhancements are valorized as ushering in a successor species. The digital age is held to proffer untold benefits and conveniences in the shape of ever more responsive and programmable 'smart' technologies and proliferation of information; but as these representations indicate, issues of access and power, especially in the context of globalization, are frequently left unaddressed within such visions. Chapter eight considers another enduring metaphor, of humans as machines. Fritz Lang's *Metropolis* weaves together associations of religion, culture and gender in a portrayal of a dehumanized, disenchanted city, and advances an account of human relations with machines that privileges the male subject, as construed in opposition to affect, the feminine and the sacred. Bruce Mazlish (1993) advances a representation of *Homo faber* – as toolmaker superseded by 'his' own creations. His thesis, of technology broaching the 'fourth discontinuity' between humans and machines, is typical of the view that the machine age will displace natural selection with technologically induced evolution. Yet the logic of erosion, dehumanization

15

and obsolescence are simply metaphors for, not predictions about, post/human development. They are themselves devices of representation by which the distinctiveness of human nature is discursively constructed. The only way of talking about human nature is via a series of paradigms or topographies plotted and interrogated through the shifting territory of the human and almost-human. This discursive turn is reflected in poststructuralist representations of identity and subjectivity, in which the 'end of the human' signals the radical interrogation of any appeal to essentialism or foundationalism. This, in turn, is being articulated in some forms of cyberpunk literature and in the work of the performance artist Stelarc. He practises a different kind of 'technology of the self' in which the materiality and performativity of embodiment suggests a model of the post/human as a constant, iterative process in interaction with and mediated by, tools and environment.

Donna Haraway's discussion of the cyborg has proved to be central to feminist discussions of the post/human. Chapter nine focuses on her distinctive approach, exemplified by what I have termed 'cyborg writing'. She uses the indeterminacy and hybridity of the cyborg – another exemplar of poststructuralist models of the subject – to articulate an ethical and political standpoint that refuses to evade the ambiguities of contemporary industrial society. Responsibility for nature, issues of women's participation in advanced technoscience and industry and the search for a renewed grounding for moral agency all emerge from Haraway's cyborg writing. Her cyborg ethic eschews the kind of transcendentalism that she equates with patriarchal religion, but the opposition between the figures of the cyborg and the goddess merely reinforces vestigial assumptions about gender and religion within a post/human technological future. So long as religious motifs continue to inform visions of the technological sublime then discourses of transcendence and re-enchantment must be directly confronted as part of an enduring symbolic of representation.

Science as salvation

Throughout the book I will be concerned to explore the implicit motifs of religion and the sacred that run throughout representations of the post/human. If the post/human destabilizes the ontological hygiene of Western modernity, then one crucial index of that is the secularism of modernity. Religion and the sacred resurface in unexpected ways within the post/human condition, such as the analogies between forms of technophilia and the world-views of hermetism and Gnosticism, discussed in Chapters four and seven. I characterize this as the re-emergence of a discourse of the re-enchantment of contemporary culture via reappropriations of ancient world-views whereby the material, immanent physical world is deemed to be but a corrupt reflection of a higher, more spiritual realm in which the face of the divine is more clearly apprehended.

In Chapter seven, therefore, I return to the rhetoric of the so-called techno-logical sublime, arguing that it rests upon an unexamined equation of 'religion' with other-worldliness, unreconstructed interiority and crude 'transcendence' (Jantzen, 1998; Carrette, 1999: 17–32). It is presumed that to be godlike is to seek mastery over creation, heedless of the fragility and interdependence of life; yet the notion of technologies transforming humanity into demiurges and gods serves to legitimate particular notions of knowledge, progress, human develop-ment, power and truth. Vital issues to do with human equity, the interdependence of the human and the non-human and the ideological nature of religion-as-transcendence go unchallenged.

There are many different ways of configuring 'transcendence' however, and I will argue that it is a vital task to expose as ideological the appeals to techno-chantment, technologies making humanity into gods; in other words, to 'polit-icise the discourses of "transcendence" in terms of the privileges and ideologies they support' (Carrette and King, 1998: 139). The first stage will be to confound the assumption that transcendence is synonymous with mastery, disembodied spirituality, fear of contingency and finitude. 'Transcendence' does not mean escaping human dependence on its environment and artefacts in pursuit of a 'technological sublime', nor does it involve insulating oneself from the instru-ments of one's own agency. Tools, artefacts and technologies are extensions and transformations of human energies, part of the activity of world-building, not a means of escape. An alternative understanding of transcendence as it informs the building and inhabiting of worlds would characterize it not as disembodied or other-worldly, but as something oppositional, visionary, undetermined – themes that will also be suggested by the unity of the material and symbolic in the work of representation, and by the radical autonomy of the non-human as portrayed in the phenomenon of monstrous *alterity*.

Whether it be the celebration of a kind of technological sublime or a concern at the loss of some spiritual integrity to humanity and nature, therefore, repres-entations of the post/human represent crucial insights into the kinds of dis-courses that will be the most evocative and authoritative for Western culture into the twenty-first century. The monstrous, the fantastic, the mythical and the almost-human serve as important bench-marks of the contest to determine whose versions of what it means to be human will prevail. This does not prevent questions about ethics and politics – about identity, participation, distribution and ultimate value – because it represents a plea for the technoscientific enter-prise to be supplemented by the cultural, literary and mythical imagination, and for the resources of popular culture, myth and narrative to inform reflections on technological progress. It is a reminder that 'the stories we live by' can be important critical tools in the task of articulating what it means to be human in a digital and biotechnological age.

Science/fiction

Part I sets out two key themes. Firstly, it establishes the inter-relationship between the institutions and practices of contemporary technoscience and the genres of science fiction, myth and literature. While often assumed to be separate and discrete discourses, concerned with 'fact' and 'fiction' respectively, I argue in Chapter one that both may be regarded as forms of representation that serve to construct the world rather than simply reflecting an *a priori* reality. In examining their significance for furnishing Western culture with influential – perhaps definitive – narratives of what it means to be human, however, it is important to have some critical tools by which such representations can be interrogated. So it is that I turn in Chapter two to the work of Michel Foucault, and to his notions of archaeological and genealogical critique. Foucault insists that 'human nature' is not ontological, but rather constructed within a network of definition, surveillance and control; and I suggest that a similar process is at work in a much older tradition of enquiry, that of teratology, or the study of monsters. Exemplary and virtuous humanity is delineated by means of its opposites, who are marked out as objects of awe and wonder by means of their aberrant nature. Such creatures on the margins of acceptable humanity – the monstrous, the 'other', the pathological, the almost-human – thus serve to delineate the fault-lines of exemplary and normative humanity.

CHAPTER ONE

Representing the post/human

[T]he very dividing line between those objects that we choose to call people and those we call machines is variable, negotiable, and tells us as much about the rights, duties, responsibilities and failings of people as it does about those of machines. The analytical point, then, has to do with the methods by which this distribution is constituted. (Law, 1991: 17)

The prospect of technologically enhanced humans, intelligent machines and modified nature places taken-for-granted assumptions about what it means to be human under increasing pressure. The erosion of clear boundaries between humans, machines and non-human nature can either be interpreted as a threat to the 'ontological hygiene' of humanity or a rendering transparent of the very constructed character of the parameters of human nature. Arguably, all forms of talk about what it means to be human – and post/human – are representations, forged within cultural contexts. Thus, the first step in gaining a critical grasp of the deeper values at work in cultural and scientific representations of humanity in relation to cybernetics, digital technologies and genetics is to recognize that both scientific discourse and popular culture operate as forms of representational practice, part of the human activities of world-building, material and symbolic.

Telling stories about technoscience

[T]he writing of declared fictions is not the only kind of storytelling going on as we deliberate about the paths we will follow in the era of biotechnology and the Human Genome Project. History and prediction, however scholarly, and however carefully built around verifiable facts about past or present, are also kinds of storytelling . . . As they are told and retold, they collide and recombine, creating new narratives, sometimes preserving old meanings, sometimes offering new ones. Together, all these stories form part of a diffuse public debate about science and technology, about what research is desirable or permissible, what applications are to be hoped for or feared, about how our society shapes and is shaped by the science it builds. (Turney, 1998: 201)

'Telling stories' about the shifting paradigms of humanity, nature and technology is always a culturally mediated activity, whether it takes place in straightforward fictitious narratives, or scientific/factual reportage, as my two following examples will illustrate.

VIRTUAL ABJECTION

The film *eXistenZ* (1999), directed by David Cronenberg, was one of a number of cinematic explorations of humanity's encounter with cutting-edge technologies to be released at the end of the 1990s.[1] The implications of digitally generated multi-sensory 'virtual reality' (VR) are explored in *eXistenZ*, the title of which refers to the name of a software program that runs a complex interactive role-play game. The film explores many dimensions of the issues I raised in the introduction, especially the blurring of boundaries between reality and illusion, technologies as dehumanizing or empowering, and the transgression of the distinction between the inside and outside of bodies. For example, the two central protagonists in *eXistenZ* articulate typically polarized attitudes of technophobia and technophilia in their contrasting attitudes towards virtual reality. The female lead, Allegra Geller, a games designer, relishes the sensual relationship she has with her software 'pod' (calling it at one stage, her 'Baby') and is a keen advocate for the potential of virtual gaming as a means of intellectual and physical enhancement. By contrast, the main male character, Ted Pikul, exemplifies an almost hysterical fear at the prospect of the encroachment of technologies upon his body.

What is particularly striking about *eXistenZ*, however, is its treatment of the effects of virtual technology on bodies as well as minds. It does not simply reiterate axioms that regard cyberspace as a disembodied realm in which users, having abandoned the 'meat machine' (Sheehan and Sosna, 1991: 139) of corporeality, experience the virtual sublime of digitally induced sensation, but recognizes morphology as a crucial site of technoscientific intervention, fabrication, enhancement, hybridity and modification. In contrast to the depiction of virtual worlds in *The Matrix*, in which digitally induced consciousness adopts an almost dream-like state, *eXistenZ* opts for a much more earthy approach. As the action traverses a number of different virtual realities, shifting from one scenario to another with giddying pace, the euphoria of transportation into graphically real parallel universes is juxtaposed with accompanying physical disorientation, for viewer and characters alike.

This attention to the physical and material impact of cyberspace extends to the portrayal of the instruments of technologies themselves. Usually, contemporary technoscience recognizes a distinction between 'hardware' (machines or robots),

---

[1] Other films to examine concepts of virtual reality, cyberspace and computer-assisted communications include *Total Recall* (1990), *Johnny Mmemonic* (1995), *Strange Days* (1995), and *The Matrix* (1999).

'software' (computer programmes) and 'wetware' (organic beings or biological materials), but *eXistenZ* transgresses these expectations. Thus the operating terminal of the virtual game is not manufactured from steel or plastic, but appears to be derived from some sort of genetically modified organic material and moulded into the shape of fleshy breasts, complete with nipple-like protuberances. This emphasis on the sensual, even sexual, nature of technologies – hardware as fetish object – is continued in the depiction of the method of connection between software console and human operator. This is in the form of an umbilical cord that slots into the spinal column (and thus the nervous system) through a biological 'port' in the small of the back. This spinal orifice resembles the human navel or, perhaps, an anal sphincter. This is not a form of technology that resides comfortably outside the skin, therefore, nor is it familiarly machinic. Rather, it comes close to a technologized polymorphous perversity, and Pikel's technophobia may be regarded ironically as the narcissistic male's resistance to the feminizing effects of penetrative and invasive technologies.

Similarly, in the unsettlingly repulsive appearance of the virtual software (with a suggestion of genetic modification, a further neat melding of technosciences) and the visceral response it evokes, Cronenberg plays with conventional representations of animals, humans and machines. By manufacturing the virtual technology of *eXistenZ* from organic tissue rather than artificial material, Cronenberg flouts conventions of technologies as externalized tools, such as keyboards, mice and remote control gadgets. And by premising the operation of the software on the absorption of operating systems into the nervous system itself, the film also transgresses the boundaries between internal and external – between bodies that end at the skin and those that are porous or penetrated. In the contrasting personalities of his main characters, and in the battles between 'realists' and 'virtualists', therefore, Cronenberg explores familiar polarizations. In the titles of the virtual games, also, he touches upon the transformative, even metaphysical, nature of virtual reality. A rival game to *eXistenZ* is entitled *TransCendenZ*.

The erosion of corporeal boundaries along with the confusion of fact and fantasy is symptomatic of widespread anxiety about the diminishment of human uniqueness in the face of new technologies. These themes are played out throughout *eXistenZ* to the final denouement itself. The audience is encouraged to identify with Ted Pikul's apprehension at the prospect of the penetration of his own body and disgust at the genetic mutations in the software factory. While *eXistenZ* is ostensibly a science-fiction thriller, it is perhaps better viewed as a horror film for the digital and biotechnological age. Indeed, Cronenberg has used similar themes – such as threats to bodily integrity, loss of control, a 'gothic' sense of lurking danger beneath the everyday – in many of his previous films.[2] The

[2] *Rabid* (1976), *Scanners* (1980), *Videodrome* (1982), *The Fly* (1986), *Dead Ringers* (1988) and *Crash* (1996).

It's official. He is almost human. And soon he could tell us why he has kept quiet so long

13·10·98

by Robin McKie

Science Editor

'WHAT a piece of work is man,' said Hamlet. 'In action, how like an angel. In apprehension, how like a god.'

But in DNA, he's just like a chimp — for humans and chimpanzees share almost 99 per cent of their genes.

In other words, only a few scraps of DNA account for all the differences separating chimps from humans, from their hairy pelts to our ability to build moon rockets. Now scientists plan to exploit these.

Researchers will meet in Chicago next month to urge US federal backing for a Human Evolution Genome Project aimed at identifying the exact pieces of DNA that separate the human genome from that of the chimp, a process that may create new med-

so close genetically, a phenomenon that has led some scientists — such as US evolutionary biologist Jared Diamond — to argue that humans should be classed as a type of chimp. Others, such as Grossman, press the case for chimpanzees to be rechristened *Homo troglodytes*.

There has already been success in identifying human-chimp DNA differences according to *Science*, which reports that scientists at the University of California in San Diego have found that chimps and humans make two slightly dissimilar forms of the molecule sialic acid. This is found on the surface of every cell in the body and is used by malaria, cholera and influenza when gaining a foothold on a cell.

'Chimpanzees are not as susceptible as humans to some of these pathogens, and

A few scraps of DNA is all that separates us from chimps. Photograph by Richard Butchins

'It's official. He is almost human.'

conventional fears of 'a human subject dismantled and demolished: a human being whose integrity is violated, a human identity whose boundaries are breached from all sides' (Hurley, 1995: 205) are only heightened by the perversely misbegotten nature of the 'technology' that encroaches, invades and dismembers.

'IT'S OFFICIAL. HE IS ALMOST HUMAN'

During the late 1990s media interest in biotechnology grew to an unprecedented level. As well as the issue of GM foods, journalists have assiduously followed the progress of the Human Genome Project, an international multimillion dollar project with the stated objective to classify the composition of all the possible configurations of DNA in the human body (F.S. Collins et al., 1998; Gruber, 1997b; Lewontin, 1993: 59–83). In October 1998 the *Observer* newspaper carried one such report on the latest advances of this enterprise. Beneath the headline 'It's official. He is almost human' and beside a comfortingly anthropomorphic portrait of a chimpanzee gazing benignly into the camera, the report explains that human beings and chimpanzees are estimated to share 99 per cent of their genetic material (see figure). 'In other words, only a few scraps of DNA account for all the differences separating chimps from humans, from their hairy pelts to our ability to build moon rockets' (McKie, 1998).

23

At the heart of this is the scientific project – elsewhere hailed as the 'Holy Grail' of scientific research (Gilbert, 1992) – responsible for delivering up an account of humanity defined in terms of its genetic composition. The article is careful not to assume that all human evolution springs from genetic imperatives, perhaps mindful that human difference from apes may be more a matter of cultural evolution, involving environment, language acquisition, social groupings, tool use and so on (Tattersall, 1998). Nevertheless, the speculation surrounding this particular narrative of human nature elevates that curious entity, 'the gene' to chief prominence. By captioning the photograph with the statement 'A few scraps of DNA is all that separates us from chimps' the report de facto effaces half a million years of evolution, tool use and language acquisition in favour of a more genetically determined explanation.

I suspect that few readers, even those unaware of the marked degree of genetic affinity suggested by the Observer report, will be surprised to discover the genetic overlap between humans and the rest of non-human nature (see also Kaku, 1998: 153). However, this particular 'scientific fact' is a relatively recent arbiter of human uniqueness – 150 years ago such a scientific depiction of biological similarity between humans and apes would have been inconceivable. Only after the publication in 1859 of Charles Darwin's On the Origin of Species did the affinity between chimpanzees and human beings become a subject of public debate, let alone commonplace observation. This point is worthy of consideration upon returning to the news report's carefully modulated representation of the affinities and differences between human and 'almost human'. The crucial actor in this is the gene. Within the discourse of the Human Genome Project and throughout contemporary molecular biology the gene occupies a number of discursive spaces simultaneously. It is a thing of nature and the very essence of life. For a biochemist it is the catalyst for the formation of essential proteins. In the bio-informatics systems that record the genes' sequences, it is a string of binary data that encodes its own particular molecular 'signature' (Cantor, 1992). In sociobiological discourse, it is the icon of destiny; and for the biotechnological corporations that stand to profit from the patenting and marketing of genetic information for medical research purposes, it is a highly lucrative commodity. The gene is a potent object of desire, and carries multiple associations. It serves as a convenient and tangible element that comes to stand vicariously for the complex mixture of environment, sociability, natural selection and biology which separates 'human' from 'almost-human'. The gene, and by association the Human Genome Project, thereby comes to represent what it means to be human; as in this report, where it is a tiny configuration of genetic material, DNA, that plays the decisive role in negotiating the mixture of curiosity and anxiety engendered by a blurring of the boundaries between 'us' and 'them'. It enables commentators to locate in the slim margin of 1 per cent the very essence of human identity and distinctiveness: their hairy pelts; our moon rockets.

Representation, rhetoric and reality

These two contemporary examples from popular entertainment and news media illustrate the potency of *representation* in shaping cultural perceptions. They are not simply objective, neutral portrayals of technoscientific innovation, but serve to construct particular discourses about what it means to be human. By 'representation' I mean any practice which expresses concepts, images, emotions and so on in symbolic form, such as language or visual imagery (Hall, 1997b: 15ff.). The practice of representation involves more than merely assigning a symbol or word to an object, as if representation were simply about attaching a label to a piece of luggage. Meaning does not reside 'in the world', nor in any inherent properties of the thing or the sign. All acts of representation depend for their effect on deploying wider discourses of meaning. Poststructuralist theory, for example, has highlighted the relationship between 'signified' and 'signifier' as governed by context and usage rather than by metaphysics (Lyon, 1999; Sarup, 1993). Representational practice is not value free, but takes place in a cultural context in which meaning inheres by virtue of convention. Members of a particular culture, moreover, exercise a degree of agency in the way they use language to represent the world. By means of 'the words we use about them, the stories we tell about them, the images of them we produce, the emotions we associate with them, the ways we classify and conceptualize them, the values we place on them' (Hall, 1997a: 3), representations also *ascribe* meaning to people, events, feelings and objects and place them in a wider discursive context of pre-existent associations.

For example, the closing years of the last century witnessed increasing public concern about the effects of genetic modification on agriculture and the food industry, both in Europe and to a lesser extent, the United States. News medias' use of 'Frankenstein Food' (as already mentioned) fuelled this. To describe produce derived from transgenic crops in such terms clearly deploys a different representational strategy to the language of 'genetically modified' or even 'genetically engineered'. Alliteration alone cannot account for the juxtaposition of the anti-hero of an English gothic novel and the activities of late twentieth-century biotechnologies and agribusinesses, but essentially this transgenic union between the material and symbolic illustrates perfectly the part played by images, associations, narratives and metaphors in shaping popular feeling and public policy.

Writing before the controversy over GM foods, Jon Turney suggested that Mary Shelley's novel has functioned as an enduring and lively resource for popular understanding of scientific innovation in genetics and biomedical science ever since its first publication in 1818. Turney calls it 'one of the most important myths of modernity . . . the governing myth of modern biology' (Turney, 1998: 3), providing Western culture with what Rehmann-Sutter terms 'interpretative patterns' (1996: 271) by which it makes sense of scientific innovation. As myth, *Frankenstein* is thus one of the narratives by which a culture constitutes   25 reality itself, a view advanced by William Paden when he argues that myths are integral to the process of what he terms 'world-building, world-shaping' (Paden, 1994: 69). A myth may narrate and construct representations of origins, relationships, means, ends and ultimate meanings; and, like *Frankenstein*, offer a mythic representation of what it means to be human.[3]

Thus particular portrayals or representations of what it means to be human, in science or in popular culture, are not simple reflections of events or objects 'in the world'. Essentially, representations also help to *build worlds*, and it is important to ask how something gets constituted through representational practices. The choice of such a term as 'Frankenstein Food' relies, consciously or not, on long-standing associations which draw genetic modification into a wider discourse of dangerous knowledge, uncontrollable nature and mad scientists. In this respect, therefore, such a representation is not simply an act of naming or describing, but a rhetorical activity. It is designed to convince, to engender particular associations and invite active responses. Representational practices serve not only to portray and report, but to legitimate, to reproduce and to normalize; or to subvert, to contradict and destabilize. Such representations are, to quote Donna Haraway, 'cultural actors' (1992c: 39) in that these portrayals participate in wider discursive networks about being human, evoking and legitimating particular values and visions. The elaboration of scientific 'facts' requires the conventions of language, texts and audience as much as fiction does. Thus 'the history of science appears as a narrative about the history of technical and social means to *produce* the facts' (Haraway, 1992b: 4, my emphasis).

Creatures of the imagination and technologized interventions into nature alike serve as embodied articulations of the scientific and cultural ambitions that designed them. It may be easier to see how an image in popular culture, or even the news media, operates in such a way. But is it really appropriate to consider that science functions as a form of representation? I want to argue that science and technology participate in building worlds of meaning as much as they are responsible for constructing cathedrals, canals and computers. Both techno-science and popular culture – literature, television and film – are influential generators of significant representations of the nature of human identity and technological futures. The disciplinary distinction between science and popular culture, or science and literature, is comparatively recent. Indeed, it may be argued, after Bruno Latour, that such a demarcation is one of the consequences

---

[3] Paden draws a distinction between myth and science: 'Scientific discourse aspires to objectivity, but religious symbols are by nature participatory, enactive, involving' (1994: 69). It should become clear that I regard scientific representations of the world as being as much narratives about origins, means and ends as religious myths. Both science and myth build representations of the world that are, albeit in different ways, binding and real. In its metaphors of naked ape, code of codes, survival of the fittest, or humans as machines, contemporary technoscience builds metaphysical or symbolic worlds of meaning as well as those of tools and artefacts.

of the categorizations of modernity, in which 'fact' is separated from 'fiction' (Latour, 1993). Yet both science and literature have common origins in culture and social history: both use generic conventions and both have authors and audiences. In short, there can be no divorce between text and context, either for literary or scientific representations of the world (Ross, 1996a).

One example of the symbiosis between science and popular culture is the extent to which science fiction since Mary Shelley's Frankenstein has drawn upon scientific practices and ideas (Aldiss, 1973). Science continued to be a rich seam of literary inspiration throughout the twentieth century as well. In the 1920s and 1930s a particularly close relationship pertained between scientific and literary practices in the publications of Hugo Gernsback (Aldiss, 1973: 208–16). Under Gernsback's patronage, science fiction, previously regarded as little more than pulp entertainment, began to play a significant role in the popular understanding of science. The editorial policy of Gernsback's Amazing Stories magazines, first published in 1926, insisted on strict scientific plausibility and a set of guidelines were composed to this effect. As cultural promoters of scientific futurism, Gernsback's titles advocated the values of progressivism and technological benevolence. Although derided as lowbrow, Gernsback's work sought to foster the values of self-improvement through science education and insisted on the public dissemination of current scientific ideas in a popular (albeit commercially lucrative) medium (Ross, 1991: 111–34; Aldiss, 1973: 201–12). Despite the naive futurist optimism and formulaic literary qualities of popular literature such as Amazing Stories, Gernsback had a keen eye for the fusion of science fiction and science realism. His characteristic vision of mechanization, mass production and social engineering helped to ease public suspicions of technological innovation. Under Gernsback's influence, science fiction as representation of a particular vision of technocratic futurism helped to legitimate the very future it predicted.

The cross-fertilization of fact and fiction, of scientific and literary invention, also informed Constance Penley's analysis of cultural representations of space travel from the era of the Cold War 'space race' to the late 1990s (Penley, 1997). Through her study of the activities of fans of the Star Trek television and film dynasty Penley shows how science fiction functions as a mythic genre in which viewers are more than passive consumers, actively participating in reworking core meanings and values. In their imaginative (and often highly subversive) reinterpretations of the world-views of studio writers and producers, Star Trek fandom and similar phenomena serve as important wellsprings of creative and critical appropriation (Penley and Ross, 1991a) (see Chapter six).

Penley's use of the term 'NASA/Trek' neatly articulates the fusion of genres which link scientific and fictional inventions. The 'race to conquer space' became central to the conduct of the Cold War. During this period NASA and its achievements became a repository for national pride, serving as a metaphor for the American Way, scientific innovation, progress. Yet NASA's dependence on federal government funding for survival has necessitated the mobilization of a relentless public-relations exercise in order to sustain public interest and enthusiasm for space travel. NASA therefore deliberately drew upon the popularity of the Star Trek series, naming one of its spacecraft after the fictional USS Enterprise and employing members of the cast in public relations roles. In its abilities to give a human (and humanist) face to outer space, to engender heroes and villains, to provide a rationale of progress and universal benefit and to promise a future in which the ambivalent power of human invention has finally been harnessed for good, Star Trek provided NASA with a ready lexicon of aspiration. This constituted an 'iconography' (Penley, 1997: 58) of exemplary achievement, individual success, enterprise, courage, national unity, progressive gender roles and race relations within a fictional medium that was readily transposed into a technoscientific context. The synthesis of aeronautic bureaucracy and syndicated space opera thus fused into 'a collectively elaborated story that weaves together science and science fiction to help write, think, and launch us into space' (Penley, 1997: 9).

As discrete entities fused into a mutually reinforcing rhetorical trajectory, 'NASA/Trek' is a powerful representation that serves to order and direct technoscientific developments into 'a common language for utopia' (Penley, 1997: 16). Scientific research allows itself to be reshaped in the image of popular culture as a means of maintaining political support for something regarded as the fulfilment of fictional dreams about space travel, galactic peace and technological progress as the enhancement rather than the diminishment of 'essential' human values. Similarly, stories about going into space become means of expressing (but possibly also displacing) hopes and anxieties about the effects of human engagement with technologies. Stephen Hawking's introduction to Larry Krauss's work of science-fiction-as-fact The Physics of Star Trek (1996), and the fact that Hawking has also made a cameo appearance in an episode of Star Trek: The Next Generation, attests again to this further blurring of disciplinary propriety. Yet arguably this is merely a continuation of a symbiosis between popular culture and scientific activity, between fictional and technological invention that informed Hugo Gernsback's science-fiction periodicals and, indeed, Mary Shelley's own literary sources (Aldiss, 1973; see Chapter three).

While critical studies of science have regarded scientific enquiry as cultural practice in which social convention shapes both empirical and epistemological procedures, scientific activity since the Scientific Revolution has portrayed itself as a value-free process of information-gathering, theory-building and application of knowledge (Jordanova, 1986a; Ross, 1991; Latour, 1993; Law, 1991). Science-as-cultural practice conceals its origins in human activity, portraying itself instead as a 'mirror of nature'. However, its theory-building is not a reporting of natural facts but a work of representation reliant on cultural and linguistic conventions.

Historically speaking, the emergence of a distinctive scientific discourse depended on certain literary conventions, conventions of representation, in which scientific texts constructed implicit understandings of the relationship between knower, knowing and known (Jordanova, 1986a). The credibility of science rested on the power of literary communication to persuade an audience of the legitimacy of scientific practice and the objectivity of that practice's findings. In order to establish an authoritative representation of the natural world, scientists constructed 'languages of nature' in which observation and empirical evidence were the foundations of scientific truth.

It would be a mistake to draw a rigid distinction between being a scientist engaged in acts of 'discovery' and being a writer . . . The two pursuits went hand in hand . . . To ask questions of nature, these have to be formulated in language, as do the answers, even before the job of constructing a narrative giving a connected account of a specific aspect of nature has begun. (Jordanova, 1986b: 110)

Bruno Latour and Donna Haraway have both located the precise historical genesis of conventions of scientific verification during the seventeenth century (Latour, 1993; Haraway, 1997). The scientist Robert Boyle (1627–91) enlisted a panel of 'modest witnesses' or ordinary laypeople to attest to the veracity of his scientific procedures. The evidence of their eyes, supposedly free of prejudice, braced by the rigours of reason, testified to the unimpeachability of Boyle's experimental method and underwrote the authority of early scientific practitioners. The space of objectivity – Boyle's 'theatre of proof' (Latour, 1993: 18) – created a clearing purged of diversions or encroachments, rendered immune to the infections of opinion or ideological belief in scientific objectivity. Boyle used language, rhetoric and representation in order to purge science of subjectivity and thus to fabricate a particular understanding of scientific truth. Yet the dominant images established in scientific texts are those of 'mapping' and 'discovering', never 'constructing' (Haraway, 1997: 31–9).

The power of science depends on the maintenance of objective, context-free criteria of verifiability and error; but it could be argued, following Michel Foucault, that this is less a matter of truth than one of power: '[T]he empirical naming and knowing of the physical world is nothing if not a culturally expressive act with fully political meanings' (Ross, 1991: 12). The qualities of reason, empiricism and objectivity are properties of science because historically they were constructed as such, and because dominant social interests defined that these categories should constitute the criteria for truth. Representations of science, even highly commodified, boundary-crossing artefacts, therefore, continue to 'naturalize' human interventions and discourses about nature, thereby veiling the 'social' legitimations that have gone into these accounts.

Questions of representation, therefore, lie at the heart of any critical appreciation of the impact of science and technology on contemporary society.

'Representation' is a means of portraying the stories that circulate concerning scientific futures, of understanding how they embody values, distribute power and insert humanity into their visions. The laboratory, the biotech corporation, the science-funding agencies are 'technical-mythic territories' (Haraway, 1992c: 42); not simply institutions for observing, classifying and discovering objective reality, but crucibles of production of the very reality we inhabit. Technologies, similarly, more than being the application of scientific epistemology, are part of the fabric of an entire symbolic and political universe.

The interpenetration of 'science' and 'technology' has thus prompted many writers to adopt the term 'technoscience' to illustrate the contention that no knowledge of the world is independent of its social context (Ross, 1991: 2–13; Haraway, 1997: 279–80). Similarly, as research is increasingly guided by commercial or political imperatives, and the social context of science – as cultural practice – is asserted, so the boundary between pure and applied, theory and practice, dissolves. Technoscience is culture because it performs the task of reflecting a world back to us and of articulating its own (increasingly definitive) version of reality: '[P]olitics involves questioning how identities are produced and taken up through practices of representation . . .' (Grossberg, 1996: 90). Technologies emerge from particular economic relations of production, bearing the marks of particular (often gendered) divisions of labour, the objectification of nature, disciplining of bodies, accumulation of capital, and pressures of commercialization. The appellation 'technoscience' thus retains a critical hold on the roots of science and technology in human labour and social relations, and ensures that conceptions of technology are not innocent of wider cultural values and aspirations:

If organisms are natural objects, it is crucial to remember that organisms are not born; they are made in world-changing technoscientific practices by particular collective actors in particular times and places . . . In its scientific embodiments as well as in other forms, nature is made, but not entirely by humans; it is a co-construction among humans and non-humans. (Haraway, 1992a: 297)

I have argued that, historically, science has been portrayed as value free, whereas it should be understood as one particular form of representational practice. This does not deny its utility or material impact, but simply ensures that its truth-claims are not placed beyond question:

To say that something is socially constructed does not make it inherently evanescent, it merely signals that we are speaking not of a (natural) given but of a (human) construct. Determining the terms under which artefacts [objects and bodies of knowledge] are constructed is a vital part of understanding them. (Jordanova, 1989: 4)

Scientific representations have also worked to bring into being other discursive categories that are crucial in shaping dominant understandings of what it means to be human in modernity. This illustrates how the authority of science produced cultural meanings through particular representational practices. Ludmilla Jordanova argues that medical and scientific knowledge of the eighteenth and nineteenth centuries rested upon representations of its own epistemological activity, and that these representations were imbued with gendered assumptions (Jordanova 1986b, 1989). Language rich in sexual metaphor was used to describe the relationship between nature and science – men's explorations of female nature forcing her to surrender her secrets – legitimating emergent scientific methods of objectification and experimentation in the name of a natural, sexualized hierarchy. Nature was portrayed as other to culture, either as unspoilt and bucolic or wild and unruly. It was gendered female. 'Science and medicine have acted as major mediators of ideas of nature, culture and gender, with verbal and visual images as the tools of that mediation' (Jordanova, 1989: 42).

I question Jordanova's use of 'mediation' if it stops short at presenting biomedical discourse as merely the displacement of contested relationships and ideas – suggesting that these things are already established and concepts are fixed, at the expense of regarding scientific representations as a constitutive site of construction. Jordanova is talking about the importance of representation for the status of nascent scientific procedures and truth-claims, which was achieved in part via the portrayal of scientific activity as authoritative. In being represented as the masterful practitioner subduing feminized nature, however, such authority rested on a gendered discourse of female nature and male culture (Jordanova, 1986b: 86–9). 'Nature' was thereby objectified, making possible a particular set of scientific practices resting on observation and objectification. Once the scientists' representation of 'nature' was accepted as objective and incontrovertible, of course, they could then justify existing gender roles by virtue of their very origins in the very same 'nature'. Arguably, this continues to the present day, where ideas about gender difference (and species uniqueness) as rooted in biology (our genes) is founded on a representation of nature as insurmountable and immutable.

The Western construction of nature as antithesis of culture, reason, mind, masculinity and science is suggestive of all kinds of hidden values about exemplary human identity articulated through the transcendence and subordination of nature. As many commentators have observed, therefore, the idea of nature is in fact a product of culture, containing 'an extraordinary amount of human history' (Williams, 1986: 70). Ideas of nature as that which existed independent of human artifice became predominant during the eighteenth century, both as something inanimate and objectified and therefore ripe for exploration, but also as a timeless, unchanging benchmark of human conduct, as in the belief that virtue was inherent in a state of nature and not dependent on the moral teachings of institutional religion (Jordanova, 1986b). 'Nature' comes to mean the essential, basic constitution of the world, embodying fundamental and immutable physical (and moral) qualities. However, at the same time it is also portrayed as objectified and malleable, a contradiction to which Ted Peters has pointed (Peters, 1997). Representations of nature thus encapsulate what Peters terms the 'Promethean' discourse (humanity as the arch-manipulator and master of nature) and that of the 'Puppet' (humanity as subject to nature's all-powerful forces).

The production of difference was also vitally important to the establishment of coherent categories. Fascination with the abnormal, the monstrous and the marginal – a topic I shall develop further in Chapter two – clearly formed one lynchpin of deliberation on the true character of 'human nature': the human defined in relation to the 'almost-human'. The fact that representations of 'nature' emerge in particular historical and cultural contexts suggests not only that the idea of nature has a history but that it may be more appropriate to think of the categories of humanity and nature as constructed in relation to each other, both materially and symbolically (Smith, 1996: 44). Thus, the idea of 'nature' had to be distinguished from that of culture and the human (Lykke and Braidotti, 1996: 247), and the history of the category 'nature' may be understood 'as a register of changing conceptions as to who qualifies, and why, for full membership of the human community' (Soper, 1995: 73).

I have already noted how the genetic affinity between humans and 'almost-human' primates calls forth speculation about the nature of human uniqueness. Throughout the twentieth century, cultural anthropology has conducted a similar debate, constructing its own representations of what it means to be human in an exercise that is as much rhetorical as purely objective or scientific (Haraway, 1992b). Donna Haraway's analysis of the various narrative worlds of twentieth-century primatology serves as an excellent case study of the value-laden nature of the scientific representation of the almost-human in the name of the rhetoric of the definitively human: 'Monkeys and apes have a privileged relation to nature and culture for western people: simians occupy the border zones between those potent mythic poles' (Haraway, 1992b: 1). Cultural preoccupations are inscribed into the scientific histories of primates, as these almost-humans have been held up as exemplars of many human attributes, thus functioning rhetorically as mirrors or extrapolations of 'us'. Representations of primates embody contending models of sociability, gender relations, competition or co-operation, selection and survival. 'Monkeys and apes have modeled a vast array of human problems and hopes' (Haraway, 1992b: 2). This is symptomatic of 'the construction of the self from the raw material of the other' (1992b: 11), of human nature not as self-evident but designated by the refracted image of the almost-human. At various times, primatologists have constructed narratives of human evolution within many different sets of metaphors which clearly carry value judgements and projections of contemporary dilemmas back into human pre-history. The 'New Physical Anthropology' of Sherwood Washburn in the 1950s enshrined the values of co-operation and universalism in the figure of 'Man the hunter/tool-maker' in the name of anti-racist, anti-eugenicist notions of human diversity and adaptability (Haraway, 1992b: 209–11).

What counts as 'nature' is, effectively, a product of the practical and theoretical activities of science and technology. Theory and practice are forms of representation that build worlds, material and symbolic. Representations of human engagement with new technologies may thus be seen as part of a longer tradition of constructing and inhabiting cultural systems of value and meaning; of building worlds that are metaphysical as well as physical. To think of representations as political or rhetorical acts is to see them as interventions in the circulation of cultural meaning. It is also to consider who and what counts as 'human' within such representations, and in particular how this is expressed in the face of technological changes which expose the fiction of human essence and which offer a proliferation of posthuman futures.

Ontological hygiene

It is important to acknowledge the symbiotic relationship between humanity and its artefacts, a blurring of agent and object, external and internal, organic and artificial. We can no longer rely, however, on such distinctions to demarcate the normatively 'human' as an enclave against the non-human. The world around is populated by creatures which defy such logic, being simultaneously natural facts and social actors. Such boundary-confusion in end-of-millennium technoscience is sharply thrown into relief when an attempt is made, for example, to categorise the 'OncoMouse™'. This is a laboratory mouse bred especially to carry a carcinogenic gene to assist research into breast cancer (Haraway, 1997, 38ff.). OncoMouse™ is simultaneously a laboratory animal, transgenic species and biotechnological commodity (Haraway, 1992c). As such, just like the 'gene' of contemporary technoscientific quest, this little rodent may occupy a variety of categorical and discursive spaces. As a creature of technologized and commercialized biology, OncoMouse™ straddles the boundaries of science, business and nature, to defy definitions that depend on their purity and discreteness (Haraway, 1997: 119).

Much of modernity, in its narratives of progress, capitalism, science, reason, human rights, has been about the demarcation of boundaries, especially in terms of production, reproduction and representation. For Bruno Latour, the ontological stability of Western modernity is underpinned by two epistemological strategies. The first, that of 'purification' creates discrete categories of species, classes and states of being. Modernity, the objective status of scientific rationality, and above all the moral claims of humanism, rest upon the construction of *alterity*. The naturalism of modernity's privileged categories must be underwritten by the effacement of its others, namely inert nature, non-humans, 'the equally strange beginning of a crossed-out God, relegated to the sidelines'

33

(Latour, 1993: 13). Thus the second symptom of Western modernity is that of 'translation', the manufacture of different types of beings into hybrids. Modernity is premised on the basis of clear taxonomic boundaries, but the very same imperative to make absolute distinctions and impermeable boundaries results also in the proliferation of 'hybrids'.

Latour's own background, in the cultural studies of science, of studying science as a social phenomenon, in which the epistemological conventions are cultural constructs rather than universal givens (Rouse, 1992: 65), means he is well placed to discern the translation of nature and culture (Latour, 1993: 3). Science and society mutually construct one another; both are simultaneously real and narrated (1993: 6). As I argued earlier, the practice of science depends on the evacuation of vitality from its objects of study, and science's elevation as the sole mediator – the representative – of nature. Simultaneously, the humanities' proper province is determined to be the things of culture, the fabricated world of the social, as opposed to the natural. Yet the work of science in transforming nature into culture also serves to dissolve the integrity of purification. At the points where the contradiction cannot be maintained – the fiction kept secret – there lives a kind of monstrosity:

[W]hen we find ourselves invaded by frozen embryos, expert systems [artificial intelligence], digital machines, sensor-equipped robots, hybrid corn, data banks, psychotropic drugs, whales outfitted with radar sounding devices, gene synthesizers, audience analyzers, and so on, when our daily newspapers display all these monsters on page after page, and when none of these chimera can be properly on the object side or on the subject side, or even in between, something has to be done. (Latour, 1993: 49–50)

Alexandra Chasin's discussion of the advent of automated teller machines (ATMs) illustrates the contradictory imperatives within contemporary experiences of computer technologies (Chasin, 1995). On the one hand, attempts are made to introduce heuristic devices and design modifications in order to make computers more 'user-friendly' and less daunting to their human operators. In the case of ATMs, these voiced machines are positioned as feminine workers in a service industry, their interaction with humans taking its cue from existing conventions of class, gender and status. In order to 'pass' successfully as 'used equipment' (Poster, 1996: 197) rather than alien technology, ATMs must be seen to reproduce – thereby to naturalize and conceal – the social conventions of deference and the division of labour within human working relations. Yet by having these roles performed by a machine, the very artifice of such conventions of service, status and differentiation is illuminated.

On the other hand, Sherry Turkle's research into people's attitudes towards computer technologies indicates the need for continual strategies of resistance to the erosion of difference between human and machine (Turkle 1984, 1991, 1995) (see also Chapter five). The convenience of user-friendliness is countered by the threat of similarity. The more capable computers become at performing human tasks, the more there is a 'romantic' reaction towards the distinctiveness of humans in terms of affectivity: 'people begin by admitting that human minds are some kind of computer and then go on to find ways to think of people as something more as well' (Turkle, 1991: 225). The definitive properties of humanity are not fixed, for people mould their accounts of difference to suit the changing capabilities of machines so that human distinctiveness is expressed reactively, in inverse terms to the perceived encroachments of machines. 'Electronics, in general, occupy a *liminal space*, challenging conventional assumptions about the differences between people and machines, as well as between living and non-living entities; such challenges necessarily entail rethinking the categories themselves, the definitions behind them' (Chasin, 1995: 93, my emphasis).

The language of liminality, suggesting cognitive or categorical disorientation in the face of change, hints at the dissonance occasioned by new technologies' blurring of secure boundaries and categories. This may suggest that accounts of human identity, and the distinctiveness between humans and non-humans (machines, animals, nature) is expressed discursively rather than existing 'in the world' as a material feature of an ontological human nature.

Attention turns, therefore, to the blurring and the boundary-making associated with representations of human nature, and a concomitant rejection of the fiction of imporous essences, even though the logic of translation and hybridity may suggest otherwise. Those who would attempt to reinstate definitive notions of human nature might then be complicit with the maintenance of what I would term the 'ontological hygiene' separating human from non-human, nature from culture, organism from machine, binary pairings whose mutual purification is complicit in the discourses of modernity. Such policing of boundaries yearns for what Mark Seltzer calls 'the principle of scarcity with respect to agency and personhood' (1992: 21), namely the exercise of exclusive definitions and watertight categories when it comes to delineating the boundaries of normative humanity. However, Latour's characterization of contemporary technological hybrids suggests that the many forms of post/humanity exhibit a somewhat greater profligacy, thereby confounding more austere regimes:

It is this notion of purity that must, in fact, be problematized. For if any progress is to be made in a politics of human or cyborg existence, heterogeneity must be taken as a given . . . Rather, it must be recognized that the world is comprised of hybrid encounters that refuse origin. Hybrid beings are what we have always been – regardless of our 'breeding'. (González, 1995: 275)

Halberstam and Livingston see the dissolution of such boundaries as manifested most visibly upon bodies, suggesting that no pure categories of race, gender or sexuality can securely define the self, and signifying a myriad of technological interventions upon the surfaces and depths of the body which blur and invade its integrity and self-sufficiency: 'The posthuman body is a technology, a screen, a projected image; it is a body under the sign of AIDS, a contaminated body, a deadly body, a techno-body; it is . . . a queer body. The human body itself is no longer part of "the family of man" but of a zoo of posthumanities' (Halberstam and Livingston, 1995: 3).

I like the juxtaposition of 'family' and 'zoo' here. They offer two useful contrasting metaphors for post/humanity and how it might be conceived. 'Family' supposes a species with a tidy genealogy, effective at naturalizing its own social conventions, counting its development in generational and temporal terms and deriving its material and reputational well-being from a purity of patrilineal descent. A 'zoo', on the other hand, may have a particular logic in the ordering of its exhibits, but makes no pretension to being a natural habitat, even though it may stress its benevolence towards its residents. Yet it also hints at a wildness lurking beneath the ordered taxonomy that is only just contained by its gates and enclosures. Any imagined state of purity and fixity is a fiction. To invite speculation on the post/human is to suspect that we are perhaps more like the 'others' than like ourselves, unavoidably contaminated by hybridity and leaky boundaries.

In conclusion, therefore, terms such as 'postbiological' (Featherstone and Burrows, 1995: 2) and 'the posthuman condition' (Halberstam and Livingston, 1995: 19) are misleading as characterizations of the human implications of twenty-first century technologies if they are understood as alluding to a transition from one axiomatic ontological state to another. I prefer the term 'post/human' (as adjective, not noun), by which I intend to expose the analytical inadequacies and ethical undesirability of adopting the prefix 'post' as a marker of evolution-ary inevitability. 'Post/human' is certainly not about the inexorable progress of 'the linear juggernaut of Enlightenment rationality' (Ross, 1991: 97) in the shape of deterministic technoscientific advances, nor about the ineluctable evolution of Homo sapiens or Homo faber into Homo cyberneticus. Rather, I will put it to work precisely in order to draw critical attention to the workings of the taxonomies that distinguish between sentient, rational creators and their artefacts – natural, virtual or machinic.

We are not concerned here with the social or historical construction of 'the person' or with the narration of the birth of modern 'self-identity'. Our concern is with the diver-sity of strategies and tactics of subjectification that have taken place and been deployed in diverse practices at different moments and in relation to different classifications and differentiations of persons. The human being, here, is not an entity with a history, but the target of a multiplicity of types of work . . . (N. Rose, 1996: 142)

The post/human should therefore be read as an interrogative marker, a critical cue, for questions concerning the authors, objects and political implica-tions of appeals to 'humanism' and 'human nature' – not to mention their

putative successors and interlocutors – and as a way of attending to the occlu-
sions and silences in such rhetoric. In a sense, the categorical instability now
apparent in talk of the post/human has been present in Western culture in other
guises for a long time; and in the next chapter I shall be arguing that the human
imagination, by giving birth to fantastic, monstrous and alien figures, has in fact
always eschewed the fiction of fixed species. Hybrids and monsters are the
vehicles through which it is possible to understand the fabricated character of all
things, by virtue of the boundaries they cross and the limits they unsettle. The
ethical and political task rests in a better understanding of the social interests and
future aspirations that lie behind these various depictions of human and post/
human futures.

The issue at stake, therefore, is not about how accurate or adequate a particular identity appears to be but how identities get formed; how definitions of
what it means to be human get produced and circulate through practices of
representation. My preferred usage of post/human denotes perhaps less a condition
(signifying a degree of fixity which, on the strength of the preceding review, I
must surely abjure) than an intervention. By intervention, I mean that talk about
representations of the post/human is an occasion for acknowledging what has
always been the case – that 'human nature' is as much a piece of human artifice
as all the other things human beings have invented.

CHAPTER TWO

The gates of difference

Dreams and beasts are two keys by which we are to find out the secrets of our nature . . . they are our test objects. (Ralph Waldo Emerson, quoted in Turkle, 1995: 22)

Myths and monsters

During the summer of 1998, the Natural History Museum in London held an exhibition entitled 'Myths and Monsters'. In an intriguing fusion of old and new technologies, all the exhibits were 'animatronic' – computer-enhanced – effigies. These displays brought to life several of the best-known mythical creatures such as 'Cyclopes', the chimera, a dragon and an extraterrestrial. The subtitle of the showing 'Unravelling the Truth' neatly encapsulated the central claims of the project, which was that what the ancients regarded as supernatural or monstrous phenomena could be explained in a perfectly rational fashion. The exhibition's accompanying catalogue reassured visitors that dragons were merely giant lizards, Cyclopes' skull the bones of a prehistoric elephant, sea-monsters simply giant squids or seals (Vadi Mecum, 1998).

While some monsters may be exaggerated versions of naturally occurring species, claimed the catalogue, other mythical beasts were entirely the work of the imagination. The Loch Ness Monster was declared to be a scientific impossibility, on the grounds that a prehistoric reptile would have been incapable of living in fresh water. Chimeras and other hybrid beasts were merely allegorical representations, although thanks to modern science the possibility might in the future present itself to engender real-life versions, in the form of transgenic species, such as the 'geep' – a cross-bred sheep-goat (Vadi Mecum, 1998).

The exhibition revealed its underlying philosophy by posing the (rhetorical) question 'Scientific fact or wild fiction?' Such a rationalistic preference for the sober facts over the magical or fantastic offers, however, a somewhat impoverished view of myth and monstrosity. By presenting such creatures as either anachronistic products of premodern superstitious ignorance or as the future products of biotechnology, the exhibition failed to account for the persistence of such myths in contemporary form, such as tales of alien and angelic visitations. The ubiquity of mythical creatures across many different historical epochs and cultures suggests they may be more than the 'wild fictions' of fevered imagination. Even in supposedly rational, secular times, they retain the power to disturb and fascinate by virtue of their liminal and ambivalent status; indeed, they may be said to perform an important function within their respective cultural imaginations and political discourses.

My contention in this chapter, therefore, will be that the study of monsters does have an enduring cultural and critical significance. Just as monsters of the past marked out the moral and topographical limits of their day, so today other similar strange and alien creatures enable us to gauge the implications of the crossing of technological boundaries. Monstrous creatures everywhere invite us to entertain what I will term 'fabulations' about the interrelationships of humans, artefacts, machines and animals in which the naturalism and inevitability of axiomatic concepts of 'human nature' are deconstructed. Monsters have a double function, therefore, simultaneously marking the boundaries between the normal and the pathological but also exposing the fragility of the very taken-for-grantedness of such categories. For their audiences – and monsters were, even before the notorious freak shows of the nineteenth and early twentieth centuries, intended to be public spectacles – monsters signalled a terrible breach in formerly inviolate categories. For such a dislocation to occur signified a heinous offence against nature (Davidson, 1991), and the horror of monsters rests in this capacity to destabilize axiomatic certitudes. In this respect, while they may excite horror, they are not, strictly speaking, representations of 'abjection', for the abject is repressed, hidden and submerged, whereas one of the functions of monsters is to be a spectacle of abnormality. Monsters are excluded and demonized, but nonetheless functionally necessary to the systems that engender and classify them. The fantastic creatures who stand at the 'gates of difference' (Cohen, 1996b: 7) are truly 'monstrous' – as in things shown and displayed – in their vital manifestation of the fault-lines of identity.

The discourse of monstrosity is therefore something which both bolsters and denaturalizes talk about what it means to be human. Insofar as teratology concerns itself not with the essence of something but the conditions of its construction, then it shows intriguing similarities with the methods of Michel Foucault in his work on the history of the human sciences. His models of 'archaeology' and 'genealogy' privilege representation, language and imagery and recognize the importance of popular and scientific discourses in the formulation of hegemonic notions of what it means to be human. Foucault argues that 'human nature' is historically conceived and emphasizes the symbiosis between the centre and peripheries of cultural discourse in constituting what counts as authoritative 'truth' about identity. This will enable me to develop the suggestion, raised in the previous chapter, that the post/human is not so much an ontological state as a taxonomic category. Foucault's work suggests that regimes of knowledge about what it means to be human are simultaneously instruments of truth and techniques of government. Such representations of the 'human' must be understood as disciplines with the powers of constitution *and* governance (Latour, 1993: 137ff.) by which the normal and the pathological are differentiated. Foucault's archaeological and genealogical readings of history may therefore be considered analogous with teratology in that both are methods 'of reading cultures from the monsters they engender' (Cohen, 1996b: 3). Those located on the boundaries of the human/almost-human – in Foucault's clinics, prisons or asylums, or in the common spectacles of monstrosity – were never superfluous but served an important function. Genealogy and teratology illustrate how categories of extremity and deviance function to delineate normative and exemplary humanity.

If an interrogation of the values currently informing representations of the post/human is to proceed, therefore, similar critical methods may usefully be deployed to those of genealogy and teratology. Just as the past is of interest to Foucault insofar as it provides him with clues (but never precedents) for the conditions under which the present may be produced, I would suggest that literature which dwells upon the fantastic, which summons up alternative universes, which imagines things differently, is engaging in a similar process of challenging the fixity of the *status quo*. In its capacity to expose the reader or viewer to the artifice of taken-for-granted worlds, science fiction uses the potential of science and technology in order to read 'against the grain of the present', but it also has affinities to utopian, gothic and fantastic flights of imagination in their capacity to circumvent established laws of space, time or identity. It is this power to displace the normal, and to expose its fragility, that places science fiction alongside teratological and genealogical enquiries.

The self made strange

Foucault's critique of post-Enlightenment modernity centres on the illusory nature of its overarching concepts 'Man' and 'history'. *The Order of Things* (1970) invokes the image of the face of 'Man' drawn in the sand gradually being eroded by the incoming tide. 'As the archaeology of our thought easily shows, man is an invention of recent date. And one perhaps nearing its end' (1970: 387). Foucault resists the Romantic ideals of 'a true humankind, a true me, or even a true madness' (McNay, 1992: 39) and follows Nietzsche in announcing the death of humanism:

Before the end of the eighteenth century, man did not exist – any more than the potency of life, the fecundity of labor, or the historical density of language. He is quite a recent creature, which the demiurge of knowledge fabricated with its own hands less than two hundred years ago: but he has grown old so quickly that it has been only too easy to imagine that it had been waiting for thousands of years in the darkness for that moment of illumination in which he would finally be known. (Foucault, 1970: 308)

The modern view of the self as one who, through the exercise of reason, activates language as a transparent medium and uses it to express a pre-eminent, autonomous agency, is redundant; indeed, it was only ever an illusion. Thus when Foucault speaks of the genesis of 'man' he is talking about the history of an idea of the human and alluding to the concepts, theories and anthropologies by which the substance of 'human nature' was articulated. The human individual is not an essential, autonomous being, but always already an observed, circumscribed creation, brought into existence via the epistemologies of the human sciences.

In the chapter entitled 'Man and his Doubles', Foucault argues that the elaborate systems of classification by which classical thought ordered the universe and brought structure to a multiple array of 'representations, identities, orders, words, natural beings, desires, and interests' (1970: 303) have all but vanished. He characterizes modernity as the stage at which representation ceased to be 'the locus of origin of living beings, needs or words' and became 'an effect, their more or less blurred counterpart in a consciousness which apprehends and constitutes them' (1970: 313). An awareness of the contingency of being, language and representation requires that modern culture is called to interrogate the very preconditions of self-knowledge: 'The human being . . . is that kind of creature whose ontology is historical. And the history of human being, therefore, requires an investigation of the intellectual and practical techniques that have comprised the instruments through which being has historically constituted itself . . .' (N. Rose, 1996: 129).

Instead of asking questions which presuppose a model of the transcendent, originating subject, the critic must learn to look for new dynamics through which the subject is constituted. This necessitates new questions:

> How, under what conditions, and in what forms can something like a subject appear in the order of discourse? What place can it occupy in each type of discourse, what functions can it assume, and by obeying what rules? In short, it is a matter of depriving the subject (or its substitute) of its role as originator, and of analyzing the subject as a variable and complex function of discourse. (Foucault, 1991c: 118)

Foucault constructs a history of ideas which emphasizes the political character of understandings of human nature. In fact, 'history of ideas' is misleading, because Foucault's emphasis is always on 'the organization of the mundane everyday practices and presuppositions that shape the conduct of human beings in particular sites and practices' (N. Rose, 1996: 128). To consider the human sciences as neutral representations of human nature under modernity is to submit to the illusion of the benevolent and ameliorative effects of enlightened and scientific institutions. Knowledge delineating what it means to be human is rather a 'regulative ideal' by which human societies are governed. Foucault traces in The Order of Things how the emergence of the human sciences, in which

41

humanity became the subject and object of knowledge, was actually shaped by material interests such as the discipline and containment of populations. Definitions of normative behaviour took place via the development of 'bio-power' that ostensibly exercised a therapeutic function, but which also served to objectify and categorize. A 'humanitarian' concern for the mentally ill, the sexual nonconformist or the criminal was articulated in regimes that disciplined as they sought to rehabilitate. Knowledge and power were allied; the supposedly 'scientific' and benevolent disciplines of modernity actually generated a language for the maladies for which they found alleviation. Madness and reason, seem-ingly each other's antitheses, are connected in that both are shaped by the same techniques of confinement and coercion; the one makes the other possible.

Knowledge, thus, is not about truth and falsehood, but is profoundly instru-mental and practical, in that it emerges in order to make particular disciplinary regimes and physical orderings possible. Knowledge is, for Foucault, never disinterested. It is always directed with a purpose: to define, to include or exclude, to exercise power. It is the grammar of power, a set of rules and procedures by which some things are valued and legitimated over others, and some things privileged and others exiled. He challenges knowledge as a set of facts or propositions and offers instead the notion of *power/knowledge*: 'a mode of surveillance, regulation, discipline' (Sarup, 1993: 67). 'Discipline' serves not only to punish but to define, and Foucault's studies of panoptic institutions and the pervasive discourse of therapeutic sexologies argue that the most effective disciplines are in fact those where self-restraint is internalized, thus obviating the exercise of overt coercion (Foucault, 1977). It is not so much that power masquerades as forms of truth and knowledge, therefore, but that truth and knowledge are, ultimately, exercises of power, a power that is ubiquitous, unceasing and wielded by a variety of agents. By attending to the 'micro-politics' of the exercise of power in specific, local contexts, the critic can understand how and to what ends power is exercised. As the conglomeration of forces which enables subjects and objects to be called into being, power 'needs to be understood as a productive network which runs through the whole social body, much more than a negative insistence [whose] function is repressive' (Foucault, 1980: 119).

In *Discipline and Punish* Foucault also attacks the notion that 'truth' emerges when power is equally distributed or suspended, that authentic knowledge is liberated when oppressive systems are abolished. There is no 'natural' or ahistorical self awaiting liberation from oppressive social structures, or a subject who exists independent of constitutive discourses. '[B]oth truth and the human subject that knows truth are not unchanging givens but are systematic, differential produc-tions within a network of power relations' (Arac, 1988: vii).

Thus from *The Order of Things* (1970) onward, Foucault is tracing what he termed an 'archaeology' of the human sciences, an excavation of the discursive resources by which 'scientific' understandings of human nature were brought into being. Foucault's model of historical criticism reads 'against the grain' of teleological history, excavating that which disrupts its unquestioned flow. 'Foucault does not aim at . . . a history of who said what and why, but a story about the web of specific sentences that were uttered, and a theory, called archaeology, of what made it possible for these sentences to be uttered . . .' (Hacking, 1986: 31).

ARCHAEOLOGY/GENEALOGY

Foucault's objective, to trace the evolution of specific 'modes of domination' (Fulkerson and Dunlap, 1998: 118) under which subjectivity and truth are manufactured, leads to his espousal of a particular critical method. Initially in the notion of 'archaeology' and subsequently in 'genealogy' Foucault seeks to excavate the contours of power/knowledge in context. Foucault's method consists of trying to identify the specific interstices of discourse and social organization and how these fuse to create particular technologies of the self. Drawing upon Nietzsche's model of 'effective' history (McNay, 1992: 14), Foucault's analysis sets out to subdue 'the kind of history that is concerned with the already given, commonly recognized "facts" or dated events, and whose task was to define the relations, of causality, antagonism, or expression, between these facts or events' (Davidson, 1986: 222) in favour of a critical approach that defies a totalizing or ameliorative telos. This characteristic approach to history is outlined in 'What is Enlightenment?' (first published in 1984):

The critical ontology of ourselves has to be considered not, certainly, as a theory, a doctrine, nor even as a permanent body of knowledge that is accumulating; it has to be conceived as an attitude, an ethos, a philosophical life in which the critique of what we are is at one and the same time the historical analysis of the limits that are imposed on us and an experiment with the possibility of going beyond them. (Foucault, 1991a: 50)

Foucault's objectives are, thus, to question what is 'natural' and, particularly in later work as the genealogical replaces the archaeological, to enquire into the actual mechanisms by which 'knowledge' produces 'normality'. The much-quoted opening to The Order of Things demonstrates the iconoclastic force of Foucault's archaeological and genealogical critique. He speaks of 'the laughter that shattered . . . all the familiar landmarks of thought . . .' when reading the categories in a 'certain Chinese encyclopedia', dividing the animals into: '(a) belonging to the Emperor, (b) embalmed, (c) tame, (d) sucking pigs, (e) sirens, (f) fabulous, (g) stray dogs, (h) included in the present classification, (i) frenzied, (j) innumerable, (k) drawn with a very fine camelhair brush, (l) et cetera, (m) having just broken the water pitcher, (n) that a long way off look like flies' (Foucault, 1970: xv).

The incongruity and illogicality of these Chinese categories emphasizes the relativism of any particular classification by which it is assumed that the world is

43

'naturally' ordered. The impact of such a perspective is twofold. Firstly, it exposes us to the unfamiliarity, the otherness of the past or another culture. Secondly, it requires us to look again at the fixity of the categories by which we are used to ordering and understanding the world. Accordingly, archaeology sets out to treat all taken-for-granted assumptions with scepticism. Foucault opens *Madness and Civilization* (1961) with an account of an eighteenth-century treatment for hysteria which involved immersing the patient in a bath until, as it was reported, the patient excreted what appeared to be diseased internal organs. This may seem ludicrous, even barbaric, to twenty-first-century readers, but Foucault's point is that such historical treatments have their own inner logic, however incongruous and misguided they may seem.

By the time of his essay 'Nietzsche, Genealogy, History' (1971) and in sub-sequent works such as *Discipline and Punish* and *The History of Sexuality Volume I*, Foucault had shifted from 'archaeological' to 'genealogical' to describe his style of critique. The difference between the two terms might be characterized thus. 'Archaeology' is an excavation of the concealed and forgotten terms on which axiomatic categories are constructed; it unearths the artifice of classification and ordering. 'Genealogy', however, attempts to trace the dynamics of power/know-ledge which give such regimes their rationale. Genealogy attempts an alternative historiography that searches out the multiple origins and familial affinities be-tween elements conventionally regarded as unrelated and ontologically separate (McNay, 1992: 37). The object of genealogical critique is thus 'to shake up habitual ways of working and thinking, to dissipate conventional familiarities, to re-evaluate rules and institutions and starting from this re-problematization, to participate in the formation of a political will' (Foucault, 1991d: 350). It involves attention to the local and specific conditions by which knowledge is possible. Genealogy reveals the artifice of that presumed to be real and natural, tracing how the familiar and the axiomatic are actually set in place via networks of power and social relations. It interrupts the flow of inevitability, and problematizes everyday terms such as self, body, sex, sanity, family and desire, revealing them as products of power rather than objects of nature. 'Genealogy studies how we constitute ourselves as human subjects. For the most part this self-constitution is not the result of active, conscious decisions, but of subliminal socialization' (Hoy, 1986: 15).

Critics who decry Foucault's reluctance to prescribe ameliorative solutions or reformist political agendas should recognize that he is insisting upon a historical method of critique that is a crucial precondition for reconstruction.

Genealogy does not try to erect shining epistemological foundations . . . it shows rather that the origin of what we take to be rational, the bearer of truth, is rooted in domina-tion, subjugation, the relationship of forces – in a word, power . . . It disturbs what is considered immobile, fragments what is thought to be unified, and shows the heteroge-neity of what is taken to be homogenous . . . (Davidson, 1986: 225)

Foucault's genealogies of subjectivity therefore pursue not generalized musings on the nature of the self, but trace the specific institutions, disciplines and practitioners of knowledge/power that are simultaneously regimes of problem-solving and moral hygiene. The unruly felon, the hysterical woman, the penitent sinner and the transgressive homosexual are not ontological states but products of the quest to make human behaviour both more intelligible and more manageable to those in power.

Foucault's linkage of power and knowledge calls critical attention to the way in which discursive practices privilege certain subjects and silence others. For this reason, Foucault has been dubbed 'the historian of alterity' (Haroontunian, 1988: 111), an acknowledgement of his claim that those excluded from and pathologized by the dominant narrative are the very conditions of its telling. Insofar as Foucault's genealogical method is designed to test 'the limit rather than the identity of culture' (Haroontunian, 1988: 117) then his intention is the excavation of alterity in order to resist the totalizing power of history that eclipses heterogeneity and homogenizes the past. 'Consequently, it is the Other, the being of nonbeing, that becomes the objective of Foucault's own discourse' (Haroontunian, 1988: 121).

However, Foucault's critical method also depends on another 'othering', that of history itself. Rendered as dissonant, illogical, it too is a thoroughgoing artifice rather than an inevitable unfolding of events toward an ultimate telos. Genealogical critique is thus always subversive and counter-cultural, reading against the grain of Hegelian models of history. It seeks to relate the 'history of the present' (Dreyfus and Rabinow, 1982: xxii) and locates the source of that which appears to be familiar at the point where it constitutes difference, dissonance and unfamiliarity. It notes the dependence of the given and the normative upon the creation of alterity and exclusion:

> In the hierarchical order of power associated with science, genealogy should be seen as a kind of attempt to emancipate historical knowledge from that subjection, to render them, that is, capable of opposition and of struggle against the coercion of a theoretical, unitary, formal and scientific discourse. (Foucault, 1980: 85)

Yet the permanent and unremitting scepticism of Foucault's archaeological/genealogical method has met with criticism. His critics argue that he suspends all normative judgements to such an extreme extent that, ultimately, he evacuates his analysis of any criteria for critique. This has been the approach of such critics as Jürgen Habermas, who accuses Foucault of relativism and nihilism (Bernstein, 1991: 152). If talk about human nature is only accessible within the parameters of discourse, then no epoch or regime can promise a system that is any more liberating or humane than another. If all human action is circumscribed by the dynamics of power/knowledge, and if all discourse is thoroughly determined, then is there any point in engaging in intellectual critique and political action? Towards the end of his life Foucault argued that he regarded his own contribution as more one of excavating crises, problems and the aporia of history than of proposing a systematic programme of reform:

[W]hat I want to do is not the history of solutions, and that's the reason why I don't accept the word *alternative*. I would like to do the genealogy of problems, of *problématiques*. My point is not that everything is bad, but that everything is dangerous, which is not exactly the same as bad. If everything is dangerous, then we always have something to do. So my position leads not to apathy but to a hyper- and pessimistic activism. (Foucault, 1991d: 343)

Thus, Foucault looks for *problématiques* in order to expose the 'dangers' in their changing forms. However, he does seem to exercise an implicit bias towards those excluded from master discourses. And even 'dangerous' is a value-laden term. But for whom, and in what ways? This still raises the question: why attend to some types of subjugated knowledge and not others?

Yet Foucault's refusal to lose his nerve in the face of humanistic protests is the substance of his critical impact. The genealogical historian is, for him, engaged in a permanent revolution, and is enjoined to expose and destabilize the givenness and complacency of axiomatic histories the better to glimpse the emergence of oppositional perspectives. Foucault's resistance to the falsity of 'permanent ahistorical standards of evaluation' commits him to a form of criticism that stresses the particularity and localism of reconstructive epistemologies, and to their fragility and contingency:

Effective history defamiliarizes the familiar in order to show how the past not only differs from the present, but also refuses to offer the present its sanction and resists assimilation to its own requirements . . . Exploring the past provides not its history, but a genealogy of contemporary phenomena by isolating their components and showing the route by which they arrive in the present. (Haroontunian, 1988: 122)

Uncomfortable as this may be for Foucault's more humanistic critics, this is the critical logic of his position. The insistence that no perspective is final, that any arrival at unitary meaning or truth-claim is a partial abuse of the play of *différance*, is itself an ethical stance. Foucault's genealogical method thus arises from 'his effort to place before the mind a constantly sounding dissonance, disruptive of our passion for historical harmony and continuity' (Bernauer, 1999: xiii), and engages 'in liberating the act of questioning' (Haber, 1994: 110). He seeks not liberation but detachment, all the more transparently to see the manoeuvrings of power: 'It's not a matter of emancipating truth from every system of power . . . but of detaching the power of truth from the forms of hegemony, social, economic, and cultural, within which it operates at the present time' (McNay, 1992: 39).

Foucault's genealogy is, effectively, not so much a methodology as 'a kind of incitement' (Dean, 1994: 19), not so much a definitive new theory as 'counter-science' (Foucault, 1970: 379).[1] Genealogical critique offers a valuable model for examining the ways in which, at the beginning of the third millennium, digital, cybernetic and genetic technologies constitute new hegemonic discourses of the self. It is now time to consider how attention to the genealogy of monsters – the objects of fear and wonder inhabiting the boundaries between human and almost-human – might serve to interrogate the dynamics of power and knowledge within contemporary representations of the post/human.

The gates of difference

You can't do anything with a monster except look at it. (Brooks, 1995: 102)

Monsters have excited popular interest and scholarly speculation throughout human history. For many cultures, the existence of any living thing that seemed to transgress the laws of nature was an object of curiosity. In antiquity, such beings were held to augur tragedy and misfortune. They were often killed and dismembered in order to find supernatural signs. Creatures who were half-human and half-animal occupied particular significance, for one way of defining what was quintessentially human was to contrast it with that which signified bestiality. Beings such as centaurs (men-horses), sirens (bird-women) and satyrs (men-goats) represented a sexualized nature which rampaged through the ordered institutions of city and family: 'centaurs crash drunkenly across the division between civilized and uncivilized behaviour' (King, 1995: 143). Tales of such creatures' defeat at the hands of heroes represented the vanquishing of baser (but undeniable) instincts and demonstrated the superiority of human virtue.

Yet already it is possible to see how such an apparently axiomatic distinction between the human and the non-human is value laden. For example, Page DuBois's study of hybrid creatures in Greek myth and civilization *Centaurs and Amazons* (1991) identifies a clear gendered logic to their depiction. The epitome of full humanity was the free man, so depictions of 'others' drew upon a symbolic hierarchy in which the Greek male citizen was defined in opposition to barbarians, women and animals. Monsters that were hybrids of women and animals embodied sexual voracity and danger, and their presence in the *polis* signified chaos and disruption. Similarly, sea-monsters such as sirens and mermaids

---

[1] While Foucault regarded the practice of constant critique as paramount, he also argued that those involved in such genealogical analysis could not shrink from realizing the political implications of their work: 'those who resist or rebel against a from of power cannot merely be content to denounce violence or criticize an institution . . . What has to be questioned is the form of rationality at stake. The criticism of power wielded over the mentally sick or mad cannot be restricted to psychiatric institutions; nor can those questioning the power to punish be content with denouncing prisons as total institutions. The question is: how are such relations of power rationalized? Asking it is the only way to avoid other institutions, with the same objectives and the same effects, from . . . taking their place.' (Foucault, 1979: 152).

have traditionally symbolized the equation of women with watery elements, but always retained a clear distinction between their proper province and that of the ship, which represented mastery of the sea (King, 1995). Thus monsters marked the 'fault-line' between appropriate social spheres as well as those between separate species.

By the Renaissance, tales of the monstrous dwelt especially upon their role as divine portents, revealing God's pleasure or wrath at human behaviour (Huet, 1993: 5–6). The strange appearance of such creatures immediately alerted all concerned to the eruption of the unexpected, or the immanence of catastrophe, as if their breach of the laws of nature was a prefiguration of a more cosmic rupture. Much of the pamphleteering of sixteenth-century Europe was especially preoccupied with monsters as supernatural phenomena and theological signs, although such literature was not above rhetoric. The religious struggles between the Roman Catholic Church and reformers such as Martin Luther or John Calvin engendered much intense polemic on both sides. Luther's pamphlet of 1523, written with Philip Melancthon and translated in 1579 into English under the title Of Two Wonderful Popish Monsters, characterizes the 'pope-ass' and 'the 'monk-calf' as prodigious allegories of the corruption of the Church of Rome. Both creatures embody specific abuses of power inherent in the life of the Church, and their appearance is designed by God to alert righteous Christians to the abominations in their midst (Davidson, 1991: 38–41).

The zenith of teratological literature in Europe occurred during the sixteenth century. Its pre-eminent text was the work of a French surgeon, Ambroise Paré, first published in France in 1573 and entitled Des monstres et des prodiges (translated as On Monsters and Marvels). Poised on the cusp of medieval scholasticism – with its indebtedness to classical philosophy – and early modern medicine, Paré combined both supernaturalist and rationalist influences. On the one hand, Paré articulated commonly held explanations for much monstrosity originating in the act of procreation, or the wrongful 'mingling of seed' between humans and animals. The ontogenesis of monsters lay in 'unnatural' acts that offended the laws of God:

There are monsters that are born with a form that is half-animal and the other [half] human, or retaining everything [about them] from animals, which are produced by sodomists and atheists who 'join together' and break out of their bounds – unnaturally – with animals, and from this are born several hideous monsters that bring great shame to those who look at them or speak of them. Yet the dishonesty lies in the deed and not in words; and it is, when it is done, a very unfortunate and abominable thing, and a great horror for a man or woman to mix with or copulate with brute animals; and as a result, some are born half-men and half-animals. (Paré, 1982: 67)

The monster is therefore the tangible, corporeal manifestation of sinful and disobedient acts. Just as God sees fit to produce creatures whose characteristics are allegories of the idolotrous and blasphemous state of the Church of Rome, so God will bring forth 'abominable' beings whose disgraceful genesis – in acts and sinful will – is physically explicit. The body of a monster demonstrates the violation of moral law by a corresponding breakdown in the laws of nature – the monstrous body thus serves as 'a moral and theological cipher' (Davidson, 1991: 48).

Yet in other ways, Paré's work attempted a more systematic and humanistic analysis. For example, he drew the distinction between those creatures resulting from defiance of divine edict ('marvels' or 'prodigies', 'things which happen that are completely against Nature'), and monsters, which are merely 'things that appear outside the course of Nature' (Paré, 1982: 3). Monsters are deformities, although their occurrence can be explained rationally, whereas marvels defy reasoned explanation and can only be the result of 'unnatural' activity. A monstrous birth might be attributed to a flaw of nature, such as an excess or deficiency of paternal 'seed' or the intervention of maternal 'imagination' (Huet, 1993). In other cases, the sheer caprice of nature alone might be held responsible: '. . . Nature has disported herself [played a trick] in order to cause the grandeur of her works to be admired . . .' (Paré, 1982: 149). Yet all this remains within a theocentric perspective, albeit while attempting to differentiate between natural and supernatural occurrences. Paré was less than consistent in his use of the terms, however, so that all instances of abnormality – including beggars faking a disability in order to gain alms – were designated by him as 'monstrous'.

Nevertheless, Paré's attempts at differentiation also, in their way, point us towards the political nature of monstrosity. His wish to 'normalize' some apparitions raises the question of the criteria used to attribute some monsters to nature or accident, while others could be regarded as signs of the wrath of God, 'to warn us of the misfortunes with which we are threatened, of some great disorder, and also that the ordinary course of Nature seemed to be twisted in [producing] such unfortunate offspring' (Paré, 1982: 6). Whose responsibility was it to 'interpret' such anomalies? Paré would, in all likelihood, have held that such apparitions – and the apportioning of blame – would have been self-evident to all persons of reason. The primacy of natural law equated what was 'natural' with divine command, so anything visibly contrary to nature was also a violation of God's commandments. This enabled the contingent character of moral prohibition to be successfully concealed; and as many critics of natural law have argued, it functions as a method of enshrining the mores of a particular historical and political culture as permanent and immutable – literally, God given. So the process by which monstrosity is ascribed – in the case of Paré, the subtle distinction between 'natural' and 'unnatural' displacements of nature – is paramount. Prodigious monstrosity, for Paré, functioned as a moral pronouncement and its horrific qualities are self-evident to all rational (and therefore upright, God-fearing, law-abiding) persons. To question this was to expose one's lack of knowledge of the fundamental laws of God. Therefore the criteria by which monstrosity or abnormality might be judged did not have to be made explicit, because no-one would have dared profess ignorance for fear of being suspected in their turn as irrational or sinful.

Thus despite monsters' (and to a lesser extent, marvels') bodily malformation the issue at stake in the phenomenon of monstrosity is not physical, but moral. The Latin and French etymological roots of the idea, monstrare and montrer respectively, indicate that a monster is something shown forth. The purpose of the monster is to reveal the divine will, specifically to embody the contours of moral order and disorder, something first argued by Augustine of Hippo in City of God (Huet, 1993: 6). The distorted form of the monster thus bore the marks of a disordered state that was the tangible sign of the transgression of natural law. Simply speaking, then, a monster was one 'whose malformed body proclaims the viciousness of his or her soul' (Heffernan, 1997: 147), and physical beauty or ugliness as an expression of a deeper, inward vice was a common theme from the Middle Ages to modernity (Baldick, 1995; Botting, 1991: 5–7).

In the work of Paré it is possible to see teratology in transition from a supernatural to a more rationalist discipline. As early modern science took root, the metaphysics of monstrosity were gradually displaced by natural philosophers' search for the systematic ordering of a secularized nature. Thus by 1620 Francis Bacon was urging fellow scientists to regard instances of monstrosity as providing valuable case studies in the workings of nature, their 'deviation' enabling greater understanding of the regularity of the paths of nature. No longer a divine portent, monstrosity was now merely an intriguing exception to the ordered workings of nature, signalling a shift of emphasis from the monster as 'transgressive phenomenon to a biological anomaly' (Curran and Graille, 1997: 2).

During the European Enlightenment, scholars such as d'Alembert, Hume and Hobbes nevertheless continued the tradition – albeit by then secularized – of regarding monstrosity as a measure against which moral rectitude could be gauged. The limits of morality, represented by the monster, indicated in inverted form the qualities of reason and benevolence by which the quintessentially human could be recognized. To depict monstrosity as anti-human was to elevate human nature as self-evidently virtuous, constitutionally incapable of acts of hatred, intentional harm and malice (Steintrager, 1997: 116). This was a necessary prerequisite to secular humanism as the proper benchmark of virtue and progress.

As the popularity of freak shows from the late seventeenth to the early twentieth centuries indicates, however, prurience may have been as much of a motive as theological edification, certainly in the popular imagination. Nevertheless, monsters attracted enduring interest from aesthetic and scientific quarters as well. Stephen Bann argues that in the eighteenth century science and popular entertainment converged, in the work of craftsmen such as Arent van Bolten, creator of imaginary monsters. Such 'creative mythmaking' (Bann, 1994: 3) did not really set out to depict real creatures but was motivated more by attempts to sharpen aesthetic and taxonomic sensibilities, the better to keep pace with scientific discoveries of ever greater diversity and complexity within nature herself.

Monstrosity thus functioned to demarcate the contours of the moral land-scape, but it also lurked on a topographical boundary, signalling the threshold between the civilized world and the unknown. From classical times onwards, the monster inhabited geographical extremities, as travellers' tales furnished the occidental imagination with visions of strange beasts whose epiphany into the ordered certainties of the known social and natural order signalled the radical reconceptualizations occasioned by new conquests and territories. Monsters also implied a moral threat from the non-Christian world, their freakish natures embodying the 'imperilling expanse' (Cohen, 1996b: 3) of uncharted realms, suggesting the urgency with which they must be converted.

Peter Ackroyd's novel *The House of Doctor Dee* is based on the exploits of the Elizabethan magus and scientist, John Dee. Dee contemplates the work of the cartographer as it discloses to him the wonders of distant shores and tells of mysterious and monstrous beings. This is typical of Dee's contemporaries' sense of wonder and curiosity at the news of marvels and prodigies brought to them by travellers. To hear of encounters with fantastic beasts was to learn that the limits of the known world were gradually being pushed back; but it was also to understand that uncharted territories held danger as well as promise:

The great world is unrolled before me, and on my desk ride the blue dolphins that love young children and the sound of musical instruments; here are the serpents, coloured in green and grey, that live six hundred years and whose heads are changed into the shape of dogs or men at will; here are sea-dragons also, marked in red, that breathe fire into the water and so cause the oceans to boil; among them glide the gryphons, the whales, and all the tribes of lesser fish that sport upon the surface of the deep . . . Who could observe such things and not wish to travel with them? (Ackroyd, 1993: 50)

As John Dee surveys his map of the world, he contemplates the gradual colonization (and taming) of the wilder extremities of the uncivilized world, now marked by the presence of marvels and monsters, but soon to be brought under the mastery of Christendom:

At the northern gate there stands Terra Septemtronalis Incognita where, it is said, dwell a tribe which hold the firestones in their mouths; in the southern extremity is conjectured to be another land of desolation. Yet not all is unknown: much has been discovered by means of good geometry and the voyages of recent years, so that the world is now marked out far beyond the confines of Bohemia or Tartary. Our navigators and cosmographers have traced the outlines of Atlantis, or the New World, where have been found the crocodile that lives for a thousand years and the quail that has the falling

51

sickness: certain provinces or domains there we have named Norumbega, Nova Francia and Mocosa, in which latter part of the world has been found the horse that weeps and sighs like a man. There also is the agopithecus, the ape-like goat whose voice is very like a man's but not articulate, sounding as if one did speak hastily with indignation or sorrow. (Ackroyd, 1993: 51)

Thus in speaking of monsters as dwelling 'at the gates of difference' (see p. 39 of this book) Jeffrey Cohen uses a most apposite spatial metaphor. It encapsulates the quintessential *alterity* of the monster as embodiment of the unknown, yet also the keeper of the portal between the same and the other.

Overall, then, monsters and marvels were seen as inviting further speculation on their origins and meaning. They were not regarded as random freaks of nature, but as portents rich in significance. Rosi Braidotti characterizes the function of monsters and prodigies as 'epistemophilic', denoting a curiosity about origins and causes (Braidotti, 1996a: 138). In both popular and scientific discourse, therefore, monsters prompted the question 'How could this have happened?' They were didactic devices, pointing people to the boundaries of good and evil and performing a 'socially regulative function' (Botting, 1995: 6). Since the purpose of monsters lay in their showing forth the contours of such a moral economy, epistemophilic discourse was necessary in order to trace the source of the moral transgression or abomination for which the monster was the evidence. In this respect, accounts of monstrous procreation offer up important clues as to key anxieties and taboos (Davidson, 1991; Braidotti, 1994a).

Feminist critics have noted the frequency with which monstrosity was linked with femininity, such that many occasions of women's transgression are held to cause irregularities of birth. One widespread example of this is the attribution of neonatal impairment to the over-abundance of maternal 'imagination' (Huet, 1993). As weaker vessels than men – after the Aristotelian world-view that characterized women as misbegotten men, a result of the malfunctioning of the reproductive process – women were unstable creatures whose physical constitution was a dangerous enigma. The permeability of the limits of the female body – giving off regular menstrual discharge, capable of bearing another life within its own body, changing shape during pregnancy – all contribute to the anxiety of the patriarchal mindset. Women's own inescapable bodiliness – couched in a string of cultural referents to female irrationality that include emotion, hormones, hysteria – puts her beyond the pale of humanity as defined in terms of the rational incorporeal subject (Braidotti, 1996a: 149). In androcentric cultures where 'human nature' is equated with characteristics privileged as male, women and monsters are 'that which is other-than . . . whatever the norm may be' (Braidotti, 1994a: 80). As Braidotti's comment illustrates, while teratologists' chief interest is in the matter of ontogenesis, the question of causation is also, inevitably, linked with that of classification. Monstrosity, femininity and deviance, as markers of *alterity*, therefore, are not stable categories but inversions of hegemonic norms. In showing forth abomination, monsters are the evidence of the crime, the symptom of the disease, the misbegotten exemplars of the 'fault-lines' by which the normal and the pathological are established. What needs to be protected is reiterated, positively and negatively, and figures who carry the undesirable qualities are made monstrous – their non-affinity with the dominant group exaggerated and distorted.

THE POLITICS OF MONSTROSITY

Monsters stand at the entrance to the unknown, acting as gatekeepers to the acceptable. Their showings-forth serve as warnings; and the horror of monsters may be sufficient to deter their audience from encroaching upon their repellent territory. However, whereas Paré's interest in monstrosity and prodigiousness showed no contempt for the afflicted or malformed creatures themselves, it is easy to see how fear and repugnance at the moral circumstances of monstrous ontogenesis might be turned into malevolence towards those designated as different, and how the ascription of monstrosity might shift from being a warning against transgression to a demonization of the very creatures themselves. Monstrosity has long been a trope for invasion, contamination, assimilation and loss of identity, the ascription of monstrous and subhuman traits serving to rationalize xenophobia and prejudice. That which is different becomes pathologized as 'monstrous' and thus inhuman, disposable and dangerous; the monster is personified as a threat to purity and homogeneity. So women, racial and sexual minorities, political radicals or those with physical or mental impairments are designated inhuman by virtue of their non-identity to the white, male reasoning able-bodied subject. Just as physical characteristics can be seen as the tangible manifestation of moral failure, the malformation of bodies can be rendered analogous to disease or malefaction in national bodies (Cohen, 1996b: 6). A slippage occurs. Difference becomes deviance (as measured against a hegemonic norm) which becomes equated with other pathologies in a process of scapegoating. The qualities of subordinate groups are unfavourably compared with those of the dominant group, and become self-fulfilling prophecies of exclusion and subjugation.

Another of the features of monstrosity is its capacity to excite contradictory yet powerful feelings. The Greek term for monster, *teras*, conveys something that is both abhorrent and attractive. The monstrous body is pure paradox, embodying contradictory states of being, or impossibilities of nature. It is both a sight of wonder – as divine portent – and loathing, as evidence of heinous sin. The monster is both awful and aweful; and insofar as the monster synthesizes taboo and desire, it further articulates its ambivalence for its creators. The establishment of normative demarcations, paradoxically, engenders the very perversions such an act seeks to annihilate. The repressed will never completely vanish. Monsters embody fearful warnings of moral transgression, therefore, yet they remind their audience of the fragility of the taboos and edicts upon which the moral order rests. For those with little stake in maintaining the *status quo* (perhaps those deemed monstrous in their own right?) they speak of liberation from convention. Monsters may serve as a warning of the folly of such transgression but may also herald new possibilities. 'Monsters signify, then, not the oppositional other safely fenced off within its own boundaries, but the otherness of possible worlds, or possible versions of ourselves, not yet realized' (Shildrick, 1996: 8).

In that respect, the monster is not merely the opposite of the same, not simply an inversion (although metaphors of doubling and mirroring do figure in many of the mythical and fictional representations of post/humanity, as I shall indicate). More precisely, monstrosity indicates the end of clear deline-ations, a chaotic mixing and miscegenation of categories that in the process of confusion indicates that their ordering is far from inevitable. It is clear that the monster is not sufficient in itself but is a spectacle, pointing to something else, congenitally a hybrid, or liminal being, and thus with no secure or stable iden-tity beyond its opposition to a pre-eminent *alter ego*. The monster, that which refuses to abide by axiomatic orderings, carries a terrible threat to expose the fragility of its defining categories and thus the fiction of normality itself. Mon-strosity is profoundly paradoxical, therefore, demonstrating both an illegitimate state of non-being and an indispensability to the very system that places the monstrous beyond the pale. Any creature that so capriciously defies the preci-sion of nature's laws requires other schemes of placement and identification than those founded on binary distinctions. New categories and ways of being have to be found:

The peculiarity of the organic monster is that s/he is both Same and Other. The monster is neither a total stranger nor completely familiar; s/he exists in an in-between zone. I would express this as a paradox: the monstrous other is both liminal and structurally central to our perception of normal human subjectivity. The monster helps us understand the paradox of 'difference' as a ubiquitous but perennially negative preoccupation. (Braidotti, 1996a: 141)

Monsters are, therefore, effectively the demonstration of the workings of *différance*. Their otherness to the norm of the human, the natural and the moral, is as that which must be repressed in order to secure the boundaries of the same. Yet at the same time, by showing forth the fault-lines of binary opposition – between human/non-human, natural/unnatural, virtue/vice – monsters bear the trace of difference that destabilizes the distinction.

Monsters are our children. They can be pushed to the farthest margins of geography and discourse, hidden away at the edges of the world and in the forbidden recesses of our mind, but they always return. And when they come back, they bring not just a fuller knowledge of our place in history and the history of knowing our place, but they bear self-knowledge, human knowledge – and a discourse all the more sacred as it arises from the outside. These monsters ask us how we perceive the world, and how we have misrepresented what we have attempted to place. They ask us to re-evaluate our cultural assumptions about race, gender, sexuality, our perceptions of difference, our tolerance towards its expression.

They ask us why we have created them. (Cohen, 1996b: 12)

As the defamiliarization of history, therefore, Foucault's genealogical critique renders modernity's axioms of subjectivity 'strange'. Teratology, similarly, shows us the 'other' to ourselves, the category against which hegemonic notions of human nature are defined. Both techniques (teratology and genealogy) serve to displace and problematize the taken-for-grantedness of identity, therefore, a method which may also be at work in various literary genres which transport their readers into alternative and fantastic worlds in order to refract and deconstruct the known. As a latter-day variant on fantastic and utopian writing, as a 'high-tech' continuation of the tradition of travellers' tales of monsters and aliens, science fiction, arguably, functions to demonstrate both the fragility of our own assumptions and the promise of alternatives.

Fabulation

As I indicated, many of Foucault's critics find his archaeological/genealogical method insufficiently politically engaged, arguing that his resistance to any kind of normative principles regarding human nature is ultimately an invitation to nihilism. However, I have tried to show that it is important to understand that Foucault regarded his work as a form of ideological critique. For Foucault, in order to act differently, it was necessary to think differently, which involved thinking about the conditions on which human nature itself is premised. The realization of alternatives for the future can only be made possible by appre-hending the strangeness of what we now take for granted, thereby subverting its inevitability. This then grants some critical space within which social critique and political action might emerge. Foucault's historical method is far from neutral or quietist; it witnesses to the significance of intellectual enquiry as an essential prolegomena to social criticism.

The study of monsters, however, is not the only literary and cultural form to adopt a manner of analysis akin to Foucault's genealogical method. Journalists, satirists and political radicals alike have used similar devices, whereby fantastic, speculative and utopian literature functions as social critique. The French philosophe Montesquieu (1689–1755) in his work The Persian Letters (1721) adopted the per-sonae of two imaginary 'Persians' in order to question many of the customs of his day. In correspondence with one another about aspects of French life, the lack of assimilation of the foreigners into the implicit values of their host culture was deployed as a showing forth of the assumptions which insiders never bothered to question. In particular, the voices of Montesquieu's correspondents were a means of attacking the religious intolerance of an absolutist Church and of making a plea for greater freedom of thought. By putting this critical thought into the mouths of outsiders, Montesquieu deflected any charges of disloyalty; and by exposing French society to comparison with other cultures was also able to argue for the contingency of his own society – a useful measure for one who favoured social reform. For Montesquieu, the foreigner functioned as a kind of *alter ego*, unaffected by prejudice or chauvanism, serving instead as a vehicle by which readers judged anew their own circumstances. Similarly, in *Gulliver's Travels* (1726) Jonathan Swift satirized the political pretensions and corruptions of his day by depicting fictional societies as refracted versions of the excesses of his desired targets (Kristeva, 1991: 105–26). Here, therefore, the alien is cast as subversive critic, rather than fearsome aberrance.

Another analogy may be drawn with utopian literature. Thomas More's *Utopia* (1516) coined the term from the Greek *eutopia*, the good place, and *outopia*, no-place (Pringle, 1997: 11–14; Kumar, 1985). Utopian literature superseded the simple travellers' tale, shifting interest from undiscovered territories and uncharted civilizations towards alternative, constructed worlds founded on polit-ical transformation. As Ruth Levitas has noted, utopian fantasy, especially in its contemporary feminist forms, functions to produce an 'estrangement' from the taken-for-grantedness of this world (Levitas, 1993: 260). In addition, it empha-sizes the provisionality of the present, and the indeterminacy of likely futures, the better to disrupt the stability of a monolithic interpretation of reality. In this respect, attention to alternative worlds is a means of exposing the taken-for-grantedness of the *status quo*; a refusal to normalize and sanction the given.

Another genre that indicates the potency of 'reading against the grain' is that of the 'fantastic'. One of the best-known works on the fantastic is Tzvetan Todorov's analysis (1975), in which he characterized literature of this kind as occasioning a profound moment of doubt and uncertainty. Faced with an appari-tion of the supernatural, the person who experiences such an event either reacts in horror (the uncanny) or wonder (the marvellous); yet the fantastic denotes that singular moment of hesitation, the suspension of taken-for-grantedness before such a decision is taken (1975: 25). The 'fantastic' according to Todorov is fundamentally a suspension of normality, designating the irruption of the 'other' in order to question the nature of what is 'real' (Todorov, 1975: 166). Todorov's emphasis on incongruity, disruption and dissonance is analagous with that of Foucault's genealogical method. The irruption of mystery and ambiguity into the ordinary necessarily calls forth adjustment, but, crucially, requires a full appreciation of the unreliability of perception, and reduces one to immobility in the face of the unknown.

Todorov himself suggested that science fiction may be one of the most striking contemporary examples of fantastic literature, a genre in which the supernatural has been replaced by 'robots, extraterrestrial beings, the whole interplanetary context' (1975: 172). Despite alien contexts and settings, the imaginary world often turns out to be a refraction of the familiar, disrupting and interrogating familiar axioms. Significantly, Mary Shelley's *Frankenstein* (1818), frequently characterized as the first work of modern science fiction (Aldiss, 1973; Haining, 1994), depends for its depictions of a fantastic world upon the use of Gothic literature, which revelled in intensity of feeling and description in order to evoke a world of disturbing uncertainty. Like its parent genre, Romance, Gothic made prolific use of antiquarian legends and settings, extremes of emotion, elements of the bizarre and the mythical, believing the immediacy of the senses to be the most vital source of experience, partly to challenge the restraints of rationalism. The unreality and hyperbole of the romantic and Gothic genres intensified their impact because they relied on nothing other than the immediacy of experience, an essential ingredient in facilitating the transition into the fantastic. To the extent to which it invoked the uncanny beneath the familiar surface, therefore, Gothic exercised a similar effect to the fantastic: 'It frees us from our inhibitions and preoccupations by drawing us entirely into its own world – a world which is never fully equivalent to our own although it must remind us of it if we are to understand it at all. It oversteps the limits by which life is normally bounded' (Beer, 1970: 3).

Similar effects may be apparent in another form of speculative fiction, the writing of so-called 'alternative' histories (Dozois and Schmidt, 1998). This genre specializes in imagining what our own world would be like if decisive events in history had taken a different path. What if the Chinese, and not the Europeans, had colonized America? What if the Axis powers, rather than the Allies, had won World War II? The skill of such writing is to make these alternative scenarios plausible within themselves; yet in the best spirit of genealogical criticism, they also expose the covert assumptions upon which our own path of history has been set. 'In one way or another, they raise the ghost of another possibility in order to investigate the groundwork of the real; they raise it in order to lay it again' (Spufford, 1996: 274). Alternative history thus enables its readers to enjoy parallel worlds for themselves, but their exposure of the premises of present reality also has the desired effect of prompting readers 'both to see and see through the structures and inevitabilities of this society' (H. Rose, 1994: 213).

Two writers, William Gibson and Bruce Sterling, possibly better known for their cyberpunk fiction, have adopted the premises of alternative history in their novel *The Difference Engine* (1990). *The Difference Engine* begins with what we know about the digital revolution of the late twentieth century, but with wit and skill it 're-enchants' (Spufford, 1996: 271) the familiar and commonplace, requiring the reader to recognize how the passage of history often depends on chance events. The alternative 'road not taken' in this case originates in the work of Charles Babbage in mechanical computation. His principles, based on the printed cards of the Jacquard loom, are now regarded as an early forerunner of the Turing machine, and hence the modern computer (Plant, 1997; Kurzweil, 1999a:

60–1). Gibson and Sterling begin with the assumption that Babbage overcame the obstacles to building a working prototype and transformed industry and culture around steam-driven data processing, thus revolutionizing Victorian society. In the novel the balance of political power has also shifted, toward a more meritocratic society, ruled by a class akin to Auguste Comte's *polytechniciens*. 'Gibson and Sterling's imaginary history returns the past to us, garishly refurbished, loudly wonderful: but those prove to be the terms on which . . . we can glimpse the past's possibility alive' (Spufford, 1996: 289).

In challenging literary critics' neglect of science fiction as a serious literary medium, Marleen Barr defends it on the very grounds that it continues the honourable tradition of evoking alternative worlds, albeit technologically rather than supernaturally engendered. For her, science fiction has facilitated important feminist experiments of the imagination. 'Most male SF writers imagine men controlling a universe once dominated by nature; most female SF writers imagine women controlling a world once dominated by men' (Barr, 1992: 4). It is something of an oversimplification to polarize science fiction in terms of gadgetry versus characterization, but Barr issues a reminder of the ideological nature of uncritical technocratic futurism (such as that which will be featured in Chapter seven) and identifies an enduring affinity between science fiction and progressive, even utopian values (see Chapter six). In order to escape from what she sees as the narrowness of conventional high-tech science fiction, therefore, Barr proposes an alternative term for the genre of feminist alternative fiction, namely 'fabulation' (1992: 4). This she defines as 'literature whose alien ingredients are concocted by the female imagination' (1992: 31). She wishes to characterize fabulation as an explicit exercise in social critique, which in this respect, of course, is reminiscent of Foucault's genealogy. Rather than allowing the incongruities of the past to disturb the complacency of the present, however, fabulation summons up 'cognitive estrangement' in a similarly iconoclastic way by deploying alternative futures to displace the familiar (Barr, 1992: 10). 'Feminist fabulation is feminist fiction that offers us a world clearly and radically discontinuous from the patriarchal world we know' (Barr, 1992: 10). Just as genealogical critique challenges fixed, unitary concepts, so fabulation rejects the assumption that any prevailing social order is the only possible system available. This is to 'canonize the monstrous' (Barr, 1992: 22), to disturb the inevitability of the norm and to restore to legitimacy alternative perspectives silenced by hegemonic world-views.

Feminist fabulation depicts patriarchal reality as the fiction, exposing it as a tenuous reality whose hegemony depends on its very ability to perpetuate ideological illusions. By creating alternative narratives through fantastic or utopian writing, fabulation enables its readers to consider the possibility that, like its fictional alternatives, the known present is also an artifice. This is entirely consistent with the spirit of the critical act of 'reading against the grain' of the axiomatic present as commended by Foucault's approach to history: '[T]hrough the creation of extravagantly fictitious worlds in which everyday reality becomes strange, there emerges the possibility of dreaming of (or having nightmares about) different and other futures, of writing new myths which will enable us to take a part in shaping our futures' (H. Rose, 1994: 209).

Post/human genealogies

At the beginning of *Discipline and Punish* Foucault identifies the nub of his interest in writing the history of the prison: 'Why? Simply because I am interested in the past? No, if one means by that writing a history of the past in terms of the present [and subscribing to the Hegelian view of a history which moves inexorably towards its apotheosis] Yes, if one means writing the history of the present' (1977: 30–1).

The strangeness and incongruity of the past preoccupied Foucault, and if read genealogically, he believed it would yield traces of alterity. 'Thus, against the background of the continuum, the monster provides an account, as though in caricature, of the genesis of differences' (Foucault, 1970: 157). I have suggested in this chapter that literature about monsters, as well as genres such as the fantastic, the Gothic, or utopian and speculative fiction, may serve to dislocate the reader from an overfamiliarity with the present. As 'fabulations' they render past, present and future made strange.

Science fiction, too, shares these preoccupations, creating alternative worlds primarily in order to refract our own back to us. By invoking the paradigm shift of estrangement, the suspension of reality, or the creation of incongruous speculations, science fiction as 'fabulation' is designed to break the hold of the *status quo*. Science fiction is also the genre, arguably, in which contemporary equivalents of teratology flourish. Where once the ancients told tales of centaurs and *djinns*, demons and angels, contemporary popular genres entertain androids, cyborgs and extraterrestrials.[2] Is the enduring popularity of such creatures a way of exploring what is fascinating and frightening, of testing the limits of our own humanity against 'the Different, the Alien, the Monstrous, the Uncanny, the Marginal and the Other' (Harper, 1995: 409)?

Foucault's rejection of the totalizing narrative of history therefore has its equivalent in a desire to avoid a universalizing account of 'human nature', by seeking to contextualize and ground understandings of what it means to be human in the interplay of specific, concrete and empirical practices, just as Foucault does to other crucial categories like madness, sexuality, criminality.

---

[2] Although the motif of an innocent stranger, misunderstood by a hostile world, is perhaps more appropriate. Such characters have frequently appeared in literature, not least in science fiction, where aliens have replaced monsters as representatives of the outcast, the marginal and the abject.

Thus he embarks on a genealogical excavation of specific depictions of the almost-human. It is crucial to think about the processes by which representations of the monstrous, the other and the alien are produced, because such images function both to iterate and to undermine understandings of normative human nature. Such a mode of critique is intended to relativize the enquiry and to open up the possibility that creatures such as monsters, aliens and others, like Foucault's exemplars of alterity, are studies in how the very boundaries of the human community get established.

Thus culture can be interpreted through its representations of monstrosity: the forbidden, the unruly, those among the secure genealogies who appear mis-begotten, those who are composed of pieces that are unharmonious assemblages of incompatible categories. Yet it is precisely by embodying such paradoxes, of incorporating seemingly discordant fragments, that these beings call into question the categories according to which a culture defines the boundaries between normal and pathological. The monster's very existence subverts *taxa* of exclusion, enclosure and containment and challenges the givenness of the supposedly 'objective' orderings at the heart of science and culture. Monsters bear witness to the power of the marginal, the Other, to demarcate the known and the unknown, the acceptable and deviant. Monsters are keepers of the boundaries between human and Other, yet by virtue of their inhabiting the 'borderlands' they promise liberation from the very strictures of binary definition. Their hybridity challenges our ontological hygiene.

Myth and science fiction have always concerned themselves with exploring the blurring and interpenetration of boundaries, so it is but a continuation of that tradition to consider those contemporary products of fictional and techno-scientific worlds who inhabit the uncharted extremities of humanity, nature and artifice. The advent of the post/human, as we have already seen, blurs many of the taken-for-granted boundaries between bodies, genders, species and representations. The beings who occupy these positions – in Donna Haraway's terms, simians, cyborgs and women – are 'monstrous' in that they destabilize evolutionary, technological and biological hierarchies that serve to privilege the rational male subject. Such fantastic figures may be, as Haraway says of the cyborg, 'simultaneously a myth and a tool, a representation and an instrument, a frozen moment and a motor of social and imaginative reality' (Haraway, quoted in Gray, Mentor and Figueroa-Sarriera, 1995: 1).

Monstrosity, genealogy and representation

Part II develops further the themes introduced in Part I, continuing to identify how understandings of what it means to be human are contested and negotiated. Two of the most enduring representations of the fabricated and artificial 'almost-human' on the margins of the normative human community are the creature in Mary Shelley's *Frankenstein*, and the figure of the Golem. Both can trace elements of their genealogical descent back to ancient myth, and yet both remain vital resources for contemporary debates about the nature of scientific enquiry, the potential and limits of human demiurgical power, and the gendered nature of technological endeavour. By virtue of their liminal status they refract and show forth the hopes, fears and anxieties surrounding humanity's engagement with its tools and technologies. Both creatures remain forever prohibited from full membership of the human community, however, and the dynamics of such exclusion demonstrate some of the prevailing criteria by which human and non-human have been differentiated.

Chapter five develops the theme of representation with specific reference to the fields of genetics and artificial intelligence, illustrating how the trajectory of technoscience embodies ideological aspirations about normative human nature. To ask 'in whose image' such exemplary visions of the post/human are made is thus also to consider what – and who – is denied a place in these projects.

CHAPTER THREE

What made Victor's creature monstrous?

The monster is not decisively human; nor, as his eventual fluency and rationality suggest, is he decisively not human. Victor inadvertently engineers not a human being but the monstrous critique of the very category. (McLane, 1996: 963)

In the previous chapter I challenged the view that representations of mythical beasts and monsters can simply be dismissed as the products of superstition or irrational fancy. Instead, I have been maintaining that the cross-fertilization of scientific discourse and speculative fiction has long played a crucial role in the formation of discourses about what it means to be human. Mythical monsters, in particular, have occupied a central position in articulating – even embodying – cultural preoccupations. By manifesting a disordered physiognomy conditioned by the transgression of a moral and theological order, they demonstrate where the true contours of that system may lie.

It seems appropriate, then, to continue my examination of representations of the post/human with the story of a monster (so-called) whose ability to shape the Western cultural imagination has been immeasurable: Mary Shelley's *Frankenstein*, first published in 1818. As the story of a living being created not by conventional reproductive means but by scientific endeavour, *Frankenstein* stands as one of the quintessential representations of the fears and hopes engendered by new technologies. Insofar as the creature at the heart of the tale is both (and neither) alive nor dead, born nor made, natural nor artificial, he confuses many of the boundaries by which normative humanity has been delineated.

As a piece of popular culture, *Frankenstein* itself inhabits Western cultural imagination as a hybrid and composite monstrosity of disparate and sometimes contradictory fragments and reinterpretations. Since its publication it has been transformed into a succession of stage plays, films, short stories and other works. At a popular level at least, therefore, *Frankenstein* has taken on a currency that is more than the sum of its parts, and this has led to many characterizing the work as a 'modern myth' (Aldiss, 1973: 23; Turney, 1998; Baldick, 1987). The significance of *Frankenstein* as cultural artefact, as phenomenon, rests not only in its many adaptations and imitators, but also in its ability to evoke strong associations, even to shape and determine the direction of subsequent understandings of science. The narratives and archetypes of Frankenstein serve as 'landscapes of fear', mapping out 'expressions of a longstanding cultural ambivalence about science in which a general recognition of the power of science is accompanied by a persistent fear of the terrible consequences that follow when scientists' obsessive, amoral curiosity leads them to trespass in forbidden areas of inquiry' (Mulkay, 1996: 157–8). The hideous creature is not only the tangible sign of the dangerous powers unleashed in his creation, but also the very agent of his creator's punishment for transgressing the laws of nature.

Although *Frankenstein* endures into the twenty-first century as a narrative embodiment of fears and anxieties surrounding the fearsome consequences of scientific experimentation (Turney, 1998), I intend in this chapter to adopt a different approach. In continuation of my analysis in Chapter two, I wish to advance a *teratological* reading of *Frankenstein*. Traditionally, monsters prompted questions about their causation: 'How could such a thing happen? Who has done this?' (Braidotti, 1996a: 139). While monstrosity is frequently attributed to consequences of flouting divine proscription or of disturbing nature's equilibrium – the hideous creature is the tangible sign of the dangerous powers thereby unleashed – I want to look in other areas for the dynamics of monstrosity. Indeed, I want to argue that monstrosity, for so long automatically assigned to the creature, may need to be distributed in different directions altogether.

The question posed by the title of this chapter 'What made Victor's creature monstrous?' thus deliberately evokes an essay on *Frankenstein* by the critic John Sutherland entitled, 'How Does Victor Make his Monsters?' (1996). Sutherland focuses on the scientific and psychosexual processes involved in the animation of Victor Frankenstein's creation, and attempts to draw conclusions as to the true moral substance of the tale. However, in asking 'what made the monster?' I am less interested in the practicalities of the creature's genesis than in attempting to trace the routes by which, within the shape of the original novel and in subsequent representations of the tale, monstrosity gets attributed. The melodrama of the strange outcast creature's quest for identity may be read as an extended interrogation of the fault-lines by which exemplary humanity is set forth.

As Mary's novel tells us, Victor was the progenitor of the thing that is known as, variously (and, according to Chris Baldick, in descending order of appearance in the narrative) 'monster', 'fiend', 'daemon', 'creature' and 'wretch' (Baldick, 1995: 48). Yet the fact that each of these terms carries a different imputation signals that, just as the device of having three narrators effects a multiplicity of perspectives within the novel itself, so there may be more than one way of characterizing the artificial being at its centre. 'Monster' and its more satanic cousins 'fiend' and 'daemon' suggest brimstone and damnation; at an early stage, having read *Paradise Lost*, the creature compares himself to Satan, the deviant offspring of his Creator. 'Wretch' suggests abjection, rather than terror. Only 'creature' could be said to be vaguely value neutral, although his lack of a more personal name hints at a deeply enigmatic quality: a lack of definitive identity.

If monstrosity is not axiomatic, but ascribed, then it functions, essentially, as a refraction of its mirror image: the delineation of quintessential humanity. Yet there is a strong suggestion that the primary monstrosity within the text may be traced to Victor Frankenstein's 'necrophilic' personality. In his ambivalent mixture of attraction and repulsion towards matters of birth, embodiment and finitude Victor embodies an obsessive attempt to cheat death and to place humanity in a position of mastery and domination over non-human nature.

Born or made?

I argued in Chapter two that the traditional distinction between humanity and monstrosity was premised on visible, corporeal qualities from which particular deductions could be made regarding a monster's illegitimate, even sacrilegious, origins. Thus the bodily imperfection of Victor's creature serves as a clue to his ontological status. His hideous being, with its physical oddities, places him beyond the pale of human culture. The physical appearance of the creature serves as an epiphany of his illicit nature; he has, as it were, to be seen to be believed. So long as he exists as an idea in Victor's mind, the prospect of his creation is heroic and exhilarating. Despite his fevered episodes of grave robbery and familiarity with vile charnel-houses, Victor is not repelled by his research, but becomes ever more intoxicated by the prospect of his own success. Yet from the moment of his creation and at the point of Victor's setting eyes on him, the being's irregular features evoke disgust and horror:

His limbs were in proportion, and I had selected his features as beautiful. Beautiful! – Great God! His yellow skin scarcely covered the work of muscles and arteries beneath; his hair was of a lustrous black, and flowing; his teeth of pearly whiteness; but these luxuriances only formed a more horrid contrast with his watery eyes, that seemed almost of the same colour as the dun white sockets in which they were set, his shrivelled complexion and straight black lips. (Shelley, 1998: 39)

The visual monstrosity serves as the rationale of the creature's marginalization by human society, even as his own voice and human sensibilities contradict such vilification. (Indeed, apart from his gargantuan size, the creature depicted in the frontispiece to the 1831 edition of the novel is not especially hideous (see figure). While hiding in the forest, the creature schools himself in language and literature and reveals himself to be far from the inhuman brute his physique might suggest. Even allowing for his somewhat naive pretensions to the life of culture, the creature emerges as sensitive and intelligent. Yet his ambition to share in the delights of human culture and learning are dashed once the de Lacey

Frontispiece to 1831 edition of *Frankenstein*.

family set eyes on him. The conflict between inner good nature and outward grotesqueness articulates the deep ambivalence of a creature torn between the liberating world of language and reason and the deterministic realm of embodiment and physicality. The creature's monstrosity stems from his being a mixture – uncomfortably and impossibly so – of speech and embodiment, of conceptual and material imagination. The mismatch between physical ugliness and intellectual beauty becomes the source of his displacement and exclusion.

By giving the being an authorial voice on a footing with the adventurer Robert Walton and the scientist Victor Frankenstein, however, Mary Shelley implicitly challenges his non-human status. The multiple narratives compound the uncertainty in the reader as to the creature's true nature. He is capable of speaking for himself, and reveals an inner world of feeling and consciousness, thereby encouraging the reader to conclude that the creature possesses an innate goodness rather than an inherently bestial nature.

As the film industry realized the dramatic potential of *Frankenstein* the myth entered a new phase. It is perhaps through Hollywood and its British counterpart Hammer that *Frankenstein* has found its widest audience. Cinema has thus been crucial in disseminating particular popular meanings of the tale. However, whereas Mary Shelley suggests that the creature's monstrosity is a result of provocation, later versions erase this by denying him speech, the crucial index of rationality and civility. From the play by Richard Brinsley Peake entitled *Presumption: Or the Fate of Frankenstein*, staged in London in 1823, to James Whale's *Frankenstein* (1931), to Kenneth Branagh's *Mary Shelley's Frankenstein* (1994), the 'monster' is silent or at best, inarticulate, a device which accentuates its brutishness. Its speechlessness casts it to the margins of culture, making it unambiguously inhuman in a way that Mary Shelley's original version resists. Subsequent elaborations on stage or screen thus leave nothing about the creature's monstrosity to the imagination, by virtue of the translation to a visual medium. The creature's visible physical imperfections are not balanced, as they are in the novel, by an independent voice, which might serve as witness to an aesthetic sensibility and rationality. The creature is thus objectified, as it were, into unequivocal ugliness, without the benefit of a compensating inner life and independent monologue. The ambivalence of monstrosity dissipates, to be replaced by pure horror.

Another possible way of tracing the attribution of monstrosity is to see *Frankenstein* as a debate about the relative weight of circumstances, environment and experience in shaping a person's character, over and against the forces of birth and inherent disposition: a debate which, even today, is held about gender difference, intelligence and disease.[1] Did the creature become monstrous because people rejected him; or was there an inherent monstrosity about him that was indelible and inevitable? Is he naturally a monster because of his hideous birth, or might he have the potential to be directed either towards civic virtue or murderous bestiality?

The contemporary reader would have recognized the criteria by which Mary Shelley's portrayal of the creature debated particular models of exemplary humanity, and whether nature or nurture constituted the more powerful influence in shaping civilized values. Mary Shelley depicts the creature living out a Roussean-like ideal of 'natural' humanity in his forest retreat, uncorrupted by social forces. He evokes parallels with late eighteenth-century fascination with the idea

---

[1] As I will argue in Chapter five, the prospect of the 'geneticization' of representations of the post/human reprises the debate about the relative weight of 'nature' versus 'nurture'.

of the noble savage or tales of feral children (Butler, 1998: xxxvii). In speculating whether the untutored and unsocialized outsider could nevertheless display appropriate and quintessential human traits, Mary Shelley's contemporaries were perhaps practising another variant of teratological enquiry: the outsider, the other, as mirroring the mores of their own civilization.

One of Mary Shelley's key influences in writing *Frankenstein* was the political ideals of her father, William Godwin, political radical, freethinker and pamphleteer. Godwin's principles articulated a commitment to the benevolence of human reason, fostered by constant aspiration towards self-improvement. Often misunderstood as a naive faith in human perfectibility, Godwin's principles were in reality driven by a vision of social change founded on the eradication of human imperfection (Baldick, 1995: 64). Godwin believed that education could civilize and serve as a social leveller (McLane, 1996). Literature and learning constituted a common human bond which distinguished humanity from the 'lower' beasts, and provided the wherewithal by which humanity could attain self-determination. Thus the creature counts himself as part of human culture by virtue of his love of literature and his affinity to those around him who seem to aspire to Godwin's world of 'intellectual and literary refinement' (McLane, 1996: 959). However, he also finds himself bewildered by the contradictions inherent in his own rejection by a society whose acceptance he so craves. 'Was man, indeed, at once so powerful, so virtuous, and magnificent, yet so vicious and base? He appeared at one time a mere scion of the evil principle and at another as all that can be conceived as noble and godlike' (Shelley, 1998: 95).

In hiding in the forest, the creature learns language and contemporary manners from the impoverished nobles, the de Lacey family. Some abandoned books – Goethe, Plutarch and Milton – furnish a rudimentary education. Hungry for human company, he engages the blind patriarch de Lacey in conversation; but the rest of the family react in horror and drive him away. As his bitterness grows, his longings for love and beauty are overpowered by feelings of alienation and become all the more frustrated. He eventually stumbles across William, the youngest member of the Frankenstein family. His initial hope that a young person, unformed and unprejudiced, might show him friendship,[2] is quickly dashed by the boy's fear; and the creature experiences his first emotions of hatred as he murders the child (Shelley, 1998: 116–17).

The being's tragedy lies in the abuse of his gentle sensibilities by the inflexibility and lack of imagination of human culture, amounting to a powerful indictment of the inhumanity of his detractors. Had the creature been willingly assimilated into human society, he could have developed a benign character.

[2] William Godwin viewed the facility for friendship as a mark of the inherent goodness of human nature.

Indeed, in his own defence he pleads that circumstances, not innate predisposition, forced him into monstrosity:

Once my fancy was soothed with dreams of virtue, of fame, and of enjoyment. Once I falsely hoped to meet with beings who, pardoning my outward form, would love me for the excellent qualities which I was capable of unfolding. I was nourished with high thoughts of honour and devotion. But now crime has degraded me beneath the meanest animal . . . I cannot believe that I am the same creature whose thoughts were once filled with sublime and transcendent visions of the beauty and the majesty of goodness. (1998: 189)

Despite his ambitions towards the civilizing power of *belles-lettres*, therefore, the creature must also come to terms with the darker aspects of human nature, suggesting that Mary Shelley wished to interrogate further – even to parody – Godwin's faith in the ameliorative potential of the humanities. Paradoxically, the creature's yearning for the privileges of human intercourse only results in his being schooled in the ways of rejection and cruelty.

In the closing testimony by Walton, Mary Shelley gives the creature a final opportunity to exonerate himself. He makes a powerful plea to be remembered as one more sinned against than sinning, and his words implicitly refute the traditional Augustinian doctrine of original sin. As a being created by a purposeful and loving Creator, rather than a capricious or random power, the creature believes himself to be capable of goodness. In a gesture part resignation, part self-sacrifice, the creature surrenders himself to the icy sea. Yet the echoes of his final peroration protest his goodness, calling others to account for his misdeeds: 'My heart was fashioned to be susceptible of love and sympathy; and when wrenched by misery to vice and hatred, it did not endure the violence of the change without torture such as you cannot even imagine' (Shelley, 1998: 188).

Some critics have attributed a Christlike character to the nature of the creature's voluntary death, seeing in him a figure 'sinned against by all humankind, yet fundamentally blameless and yet quite willing to die as a sacrifice' (Oates, 1984: 547). The creature's death seems more despairing than heroic or sacrificial, however; for while the suggestion of the creature as wronged victim is powerful, the text also suggests that a corruption of character does indeed take place. The creature may begin his life untouched by sin, but the actions of others in ascribing to him an excluded and monstrous identity become a self-fulfilling prophecy. Like that of Satan in *Paradise Lost*, however, the creature's fall from grace must be genuine in order for readers to experience the full weight of his tragedy; hence the trail of death and destruction. The creature is pitiful, but goes to his solitary, bitter death convinced of his estrangement and damnation. Far from being redemptive, the creature's suffering is the sign of his *dehumanization*.

Naming the beast

Despite attempts to cross the boundaries between himself and the rest of the human race, and despite the eloquence of his speech, therefore, the creature's overtures to human society always end in rebuttals. The creature longs to be heard as a rational being, not to be captured in the objectifying and 'pitiless gaze of the other only as the witness to his inescapable monstrosity' (Heffernan, 1997: 158). Whatever his pretensions, the creature is imprisoned by the exclusion of civil society in which the outward marks of difference are absolutized. Reminiscent of racial or sexual prejudice that causes physical traits to determine social status, Peter Brooks notes the dichotomy between language and corporeality, which stubbornly insists on the creature's otherness. In Lacanian terms, the creature longs to be accommodated within the economy of the symbolic, yet the physical manifestations of difference refuse to yield in order for him to be, as it were, acculturated (Brooks, 1995). In this respect, the creature is placed in an objectified role, as the object of external power, as one who is forbidden from gaining access to the compensations of full subjectivity. The creature's own entrapment in his body echoes the 'literal monstrosity many women are taught to see as characteristic of their own bodies' (Gilbert and Gubar, 1979: 240). Like many oppressed and dehumanized peoples he is 'enclosed' in difference.

The monster cannot break out of the exile imposed upon him by the 'ontological hygiene' that casts him as inhuman purely by virtue of his illicit beginnings, regardless of his achievements or aspirations. This is evident in his inability to place himself within a genealogy, or to claim a home town or country: 'he cannot appeal to familial, political or other territorial categories which would provide him with techniques of authentication and remembrance' (McLane, 1996: 967). Whereas Victor's 'natural' state is civic (he begins his story to Walton with the words 'I am by birth a Genevese, and my family is one of the most distinguished of the republic'), familial, propertied and educated, the creature's life is solitary, impoverished and uncivilized, but none of these terms compares to the being's lack of name. His namelessness, alluded to earlier, reinforces his problematic status, signifying the absence of a personal – or 'Christian' – name (and the usual intimacies that go with it) and of a patronymic by which an individual might establish his or her social and economic position. The creature thus lacks the essentials of 'high and unsullied descent united with riches' (Shelley, 1998: 96) that would protect him from the vicissitudes of nature and guarantee him a place within the social hierarchy.

In the absence of familial connections to legitimate his inclusion in civil society, the creature has to rely on the fantasies about his origins and his true destiny derived from the very texts he uses to educate himself. Thus he styles himself after Adam (in his uniqueness and solitude) and Satan (in his envy and self-hatred) in *Paradise Lost*. He is dependent on a system of prescriptions and norms for virtue parroted from Goethe's *The Sorrows of Werther* or Plutarch's *Lives* (Shelley, 1998: 103–4). He is ignorant of the circumstances of his creation until he reads Victor's journal, a surrogate for the relationship that never was, and never will be, with his creator (1998: 105).

The creature is thus trapped in a ceaseless circulation of unnamable and deferred qualities, a lack of definitive and self-directed identity. His attempts to gain the status of a speaking subject who is not caught in the endless loop of others' significations are futile. The autodidact and aspiring gentleman of letters hopes to prove himself worthy of human intercourse; but his self-construction has no meaning beyond the texts he has gleaned. These have substance 'only as they are verbal matrices naming an absence and only as they hold out different promises of meaning with no foundations aside from linguistic assumptions' (Hogle, 1995: 224). This view departs from a humanistic reading in that it confronts the possibility of the creature's monstrosity lying in the very nature of his genesis. Formed from fragments of mortality, the monster is only negativity and absence, emerging from the void of death – the dark foulness of the charnel-houses – and stripped of the privileges of familial or generational sources. While Victor and his creature may be twinned in many respects, they remain on opposite sides of the human/almost-human, culture/nature, civil/pastoral dichotomies. The monster is forever stranded on the wrong side of human subjectivity, despite all his efforts at self-improvement through intellectual pursuits. 'In entertaining humanist fantasies, the monster forgets his corporeally and nominally indeterminate status' (McLane, 1996: 975).

Can anything end the creature's solitude? The request for a mate, another of his own kind, signals his capitulation, in the face of his social and ontological exclusion, to the realization of his difference. Victor's skill in the natural sciences – which ruptured the boundaries of species in the first place – can deliver another of the same species; yet Victor refuses, realizing the implications of providing a mate. If the two beings were to reproduce, Victor's original dream of being the progenitor of a new race would indeed come true, but at the cost of founding a parallel race whose continued, collective existence might well in time transform the creature from a monstrous singleton into a collective. Victor associates the creation of a second 'thinking and reasoning animal' with the threat to humankind and to the axiomatic integrity of his species' humanity. His bonds to his own creation are ultimately weaker than his loyalty to his own species and so he aborts his half-finished female being without a qualm. In so doing, Victor reiterates his preference for a 'reified category for human fellowship conceived over and against the monstrous alternative' (McLane, 1996: 982), or the ontological hygiene of species differentiation, despite his own experimentation on the boundaries of life and death.

Revolting monsters

While representations of monstrosity in *Frankenstein* may reflect debates about individual character formation, they are also suggestive of political debates at the beginning of the nineteenth century about the inclusion of an entire social class into the collective civil community. Political enfranchisement and the true extent of the vision of the 'Rights of Man', along with Revolution, or fear of it, were ubiquitous themes in tracts, journals and pamphlets of the day (Baldick, 1995). The spectre of the French Revolution a generation earlier had also haunted the public presses. The fear that unrest would spread like a contagion over the English Channel was a device much used in conservative journalism from the 1790s. Similarly, the years following the cessation of the Napoleonic Wars in Britain were characterized by a series of labour disputes, riots, murders and trials, and it was common to portray those involved as a destructive and godless mob, bent on the violent overthrow of Parliament, Church, Monarchy and private property.

Just as physical monsters might bear witness to a moral disorder, a rupture in the natural balance of creation and procreation, so too the idea of monstrosity had been applied to threats to the body politic from the seventeenth century (Baldick, 1995: 52). The threat of bodily dismemberment, occasioned by unlawful interruption of the divine right of kings, represented an attempt to manufacture an artificial body politic, a monstrous departure from divinely ordained rule. The idea of monstrosity was extrapolated from that of an offence against nature into a political metaphor, reaching its apotheosis in Edmund Burke's *Reflections on the Revolution in France* (1790). Burke argued from a conservative standpoint that derived from a Hobbesian social model of organic unity whereby it is understood that a stable constitution rests on a delicate, and divinely sanctioned, system of interdependent elements. Sudden and drastic social change, such as had occurred in France, represented a tearing asunder of natural elements, producing a monstrous abomination of dysfunctional institutions.[3]

The insistence of the *philosophes* that principles of rationality, and not tradition or superstition, should dominate human governance, leads Burke to characterize their politicial incursions as an unnatural and 'abominable perversion' (Burke, 1983: 161), condemning their hubris in the language of ghoulish foreboding. Out of their sorcery looms a social experiment laden with monstrosity, transgression and disaster. He speaks of the victims of the mob, envisaging the scene as one 'swimming in blood, polluted by massacre and strewed with scattered

---

[3] For a graphic description of political sedition understood as threat to the body politic, see Michel Foucault's discussion of the public torture of a regicide in eighteenth-century France. The punishment is made to fit the crime as the accused is drawn, quartered and burned, the retribution for his crime being a symbolic re-enactment of his attempt to dismember royal authority (Foucault, 1977: 3–6).

limbs and mutilated carcasses' (Burke, 1983: 164). For Burke, the idealism and overweening ambition of the Revolution was doomed to unleash unforeseen powers and result in a descent to anarchy. Burke represents a particular under-standing of the social order, in which authority and a shared moral consensus form essential bonds which protect humanity from its worst instincts.

Burke's polemic against the Revolution attracted a torrent of criticism; and an alternative, more radical perspective may be discerned in the views of Burke's opponents. In *The Rights of Man* (1791) Tom Paine challenged Burke's use of the metaphor of monstrosity. The offence against nature for Paine lay not in the actions of the reformers but in the 'artificial exaggerations of wealth, rank, and privilege' (Baldick, 1995: 57) on the part of the aristocracy. Whereas Burke saw democracy as monstrous, an aberration of the natural state of hierarchy and order, Paine regarded the artificial distinctions of class and estate as unnatural.

Similarly, Mary Shelley's own parents, William Godwin and Mary Wollstonecraft, were also active in the debate, siding with Paine against Burke. In Godwin's *Enquiry concerning Political Justice* (1793) feudalism and aristocratic rule are described as a 'ferocious monster' against which the gradual but inevitable encroachments of a system founded on reason and enlightenment will prove unstoppable. In her *Historical and Moral View of the Origins and Progress of the French Revolution* (1794) Wollstonecraft argues that it is the ruling class's neglect of duty and their pursuit of indulgent pleasures that are hubristic, not the agitations of the mob.

Yet even those who supported the legitimacy of the Revolution from the other side of the English Channel – not to mention many reformers among the French bourgeoisie, including Voltaire – were fearful of the violence that was unleashed. With the rational and scientific new Republic collapsing into Jacobin fanaticism, radicals such as Godwin and Wollstonecraft asked themselves how a system conceived in the spirit of human emancipation and progress could be foundering. How could scientific study conceived in a spirit of altruism become so basely corrupt? What if the chaotic outcome of a Revolution conceived as the bloom-ing of rational self-determination only served to expose the shadow side of enlightenment?

One of the first references to *Frankenstein* in English literature joins the debate on this very point. In the face of social injustices, are there any grounds to justify seditious or violent actions? Elizabeth Gaskell's *Mary Barton*, published in 1848, reflects the turmoil of a year of revolution throughout continental Europe which had rekindled anxiety in Britain at the prospect of similar unrest. So far as British commentators were concerned, Chartism represented the monstrous threat to the *status quo* at the time. Seeking to defend John Barton's Chartist and socialist sympathies, Gaskell asks whether the actions of revolutionaries can be condemned when the extremities of poverty drive such people to desperation. While some sections of the working classes may resort to opium, rather than communism, they are not driven to such measures by innate vice, but by the severity of their deprivation:

The actions of the uneducated seem to me typified in those of Frankenstein, that monster of many human qualities, ungifted with a soul, a knowledge of the difference between good and evil.

The people rise up to life; they irritate us, they terrify us, and we become their enemies. Then, in the sorrowful moment of our triumphant power, their eyes gaze on us with a mute reproach. Why have we made them what they are; a powerful monster, yet without the inner means for peace and happiness? (Gaskell, 1970: 219–20)

Gaskell may have coined the habit of mistakenly referring to the creature as 'Frankenstein', and not Victor, but her argument nevertheless epitomizes what was a recurrent sentiment on the part of the liberal intelligentsia and reforming middle classes throughout the early Victorian period: if the common people are to be labelled monstrous, then those responsible for the creation of the conditions that engendered them must also be held to account.

Was Victor a 'mad scientist'?

Mary Shelley's own initial impetus to create Frankenstein may thus have been fuelled as much by debates about character formation, civic virtue and political insurrection as fears about the catastrophic excesses of science – at least, in its earliest form as short story and novella. By 1823, however, when Frankenstein was first dramatized, Mary Shelley was already beginning to make the scientific themes more explicit, giving a more overtly moralistic tenor to her brainchild. 'The striking moral exhibited in this story is the fatal consequence of that presumption which attempts to penetrate, beyond prescribed depths, into the mysteries of nature' she wrote (O'Flinn, 1995: 31). By 1831 and in the revised preface to the second edition this tone is even more apparent: 'supremely frightful would be the effect of any human endeavour to mock the stupendous mechanism of the Creator of the world'. The theme of forbidden knowledge is now accompanied by a clear theological subtext. Yet to recontextualize Mary Shelley's novel is to rediscover a subtlety of understanding towards the emergent physical sciences that is worthy of further consideration.

[I]t would be a mistake to call Frankenstein a pioneer work of science fiction. Its author knew something of Sir Humphrey Davy's chemistry, Erasmus Darwin's botany, and, perhaps, Galvini's physics, but little of this got into her book. Frankenstein's chemistry is switched-on magic, souped-up alchemy, the electrification of Agrippa and Paracelsus . . . He is a criminal magician who employs up-to-date tools. Moreover, the technological plausibility that is essential to science fiction is not even pretended at here. (James Rieger, introduction to Frankenstein, quoted in Florescu, 1976: 241)

James Rieger's comments are typical of critics who disavow any scientific context to Frankenstein, choosing instead to characterize the novel as the product of a febrile preoccupation with occultist and supernatural themes. Rieger's perspective may, of course, be in part a reaction against the axiom which fuels so much of the 'mythical' status of *Frankenstein*, that Mary Shelley wrote it as prophetic warning against the excesses of scientific licence. Certainly, any consideration of the multiple histories of *Frankenstein* – especially in the light of Jon Turney's thesis – ought to take the widely held assumption into consideration that *Frankenstein* is about a 'mad scientist'. On the other hand, neither interpretation does justice to Mary Shelley's evident knowledge and interest in the emergent discipline of natural science, and to her attempts to represent some of the contemporary debates about the proper conduct of scientific procedures. She was motivated more by questions of experimental method than of scientific madness, granting further insights into the genealogy of monstrosity in *Frankenstein*.

Like James Rieger, John Sutherland believes that *Frankenstein* must be read more as a work of occult or gothic literature than as an early example of science fiction shaped with any degree of credibility by the scientific practices of its day. *Frankenstein* is, after all, written in the style of a popular literary genre of the time, that of the Gothic novel. Gothic emphasizes the uncanny lurking beneath the prosaic and everyday, the darkness and horror at the heart of beauty, which at first acquaintance could be regarded as the antithesis of the Enlightenment's commitment to empiricism and rationalism. Sutherland argues that later readers of the novel have allowed themselves to be swayed by the extent to which subsequent cinematic adaptations indulge in extravagant representation of the scientific mechanics, but that Mary Shelley's own account of the process is closer to the magical incantations of the creators of the golem (see Chapter four) than 'the surgery, transplants, and electrical apparatus of cinematic folklore' (Sutherland, 1996: 31). Sutherland adjudicates, 'Shelley's Frankenstein is no "scientist", whether mad or sane, but an Enlightenment *philosophe*' (Sutherland, 1996: 25). Yet a *philosophe* would surely have abjured the ghostly overtones of gothic literature in favour of the rationalism of scientific method. Thus Sutherland's assumption that Mary Shelley would have been ignorant of natural philosophy runs counter to a mass of evidence to the contrary. In assuming that women's writing can only reflect emotional and domestic concerns, Sutherland misjudges Mary Shelley's informed understanding of scientific controversy – an awareness, indeed, that was far more considered than mere anti-scientific polemic.

Legitimately enough, Sutherland's assessment is based on the observation that the progress of the researches is not given in great empirical detail (Sutherland, 1996: 33–4). However, this may be related to the problematic relationship between a textual narrative and its cinematic adaptation. Film versions have to find ways of depicting the manufacture of the creature vividly in a visual medium, whereas the text itself can concentrate instead on the emotional intensity of Victor's activities. This is not accidental, for the influence of the Romantic and Gothic movements upon Mary Shelley led her to place an emphasis on the passions as a more authentic means of representing human motivation and the creative process. The elements of Romanticism in *Frankenstein* – the emphasis upon the emotions and inner thoughts of both Victor and his creature – reflect a strongly humanist bent, in which the human subject, and not supernaturalism or divine agency, is placed at the heart of their world-views. Thus her emphasis on the human – a subject both passionate and rational – fully reflects both a rationalism in which the mind, unfettered by superstition, pursues knowledge and a romanticism in which the immediacy of the passions, unmediated by external authority, seeks truth. By creating an artificial being against this backdrop, Mary Shelley broke with the more ancient conventions of automata, golems and animated statues, thus establishing herself as a writer at the very heart of modernity. Unlike Goethe's *Faust*, published around the same time, Frankenstein is not brought low by his involvement in demonic or supernatural forces, nor is he punished by the Fates or the gods, as in earlier myths.

While she may have chosen to opt for what Sutherland terms 'Gothic fuzz' (1996: 29) at the climax of her tale, therefore, Mary Shelley also intended *Frankenstein* to explore serious issues of natural philosophy in the context of the scientific debates of the time. As the daughter of a famously rationalist father, Mary Shelley was well versed in the scientific innovations of her day and eager to engage with new discoveries and new methods. As a young girl she frequently accompanied her father to scientific lectures and demonstrations. Among those scientific discoveries with which Mary was familiar would have been Benjamin Franklin's (1706–90) experiments proving that lightning was electrical in nature.[4] William Godwin was an acquaintance of Sir Humphrey Davy (1778–1829) and Mary read Davy's *Elements of Chemical Philosophy* (1812), also about electricity. The experiments of Luigi Galvini (1723–98), which seemingly made frogs' legs jerk and convulse by means of 'galvinism' or electrical current, were also known to her (Mellor, 1995: 124–5). She includes the use of electricity in the novel to parody the thesis that life is animated by metaphysical spirit or soul, rather than being a product of the emergent properties of materialist life (Butler, 1998: xxx–xxxiii). Far from representing a lapse into Gothic emotivism, Mary Shelley's focus on Victor's personality was central to her depiction of the disastrous implications of his flawed scientific method. By setting *Frankenstein* in her own day, and not in antiquity – perhaps more typical of the Gothic genre – Mary Shelley was aligning herself with the protagonists in some of the most controversial intellectual debates of her day (Butler, 1998: xv–xxxiii).

Mary Shelley also owed a considerable intellectual debt to the physician and engineer Erasmus Darwin (1731–1802), grandfather of Charles. He receives

---

[4] In Kenneth Branagh's *Mary Shelley's Frankenstein* (1994), Victor, Elizabeth and his family are seen building an elementary lightning conductor out of kites. Mary and Percy Shelley flew balloons on Lake Léman during their stay with Byron in 1816.

acknowledgement in the preface to the 1818 edition and in the new introduction to the 1831 revision. He may have introduced her to legends and practical examples of automata and artificial machines, but as a freethinker and atheist, he would also have advanced materialist and not supernatural views on the origins of life. The predominant model for Enlightenment science was founded on a belief in the natural equilibrium of nature. The self-regulated order of living species could be observed and noted, and taken to be exemplary of a similar balance and harmony in human societies. Erasmus Darwin himself had begun to put forward some tentative views on the nature of evolution in his *Zoonomia* (1794). The taxonomic model of natural philosophy – indeed, the suggestion that nature herself, by dint of organic, gradualist change, was responsible for the development of species – precluded the rude interventionism of the alternative approach, styled after Francis Bacon. After Bacon, natural philosophers sought dominion over the created order: 'I am come in very truth leading to you Nature with all her children to bind her to your service and make her your slave' (Mellor, 1995: 107). Under the tutelage of Professor Waldman, Victor learns how the ancient philosophers' flirtations with the origins of life have been surpassed by the expertise of the natural sciences, especially chemistry.[5] This model of knowing is reminiscent of a Baconian world-view, as Waldman describes the activities of the heroes of scientific endeavour thus:

They penetrate into the recesses of nature and show how she works in her hiding-places. They ascend into the heavens [rather like the risen Christ]; they have discovered how the blood circulates and the nature of the air we breathe. They have acquired new and almost unlimited powers; they can command the thunders of heaven, mimic the earthquake, and even mock the invisible world with its own shadows. (Shelley, 1998: 30–1)

This echoes philosophical perspectives on technology where humanity harnesses elemental powers, usurping God via the human assumption of nature's destiny; a view not only Christian but pagan (Cooper, 1995: 9). Victor's obsessive interest, framed in the language of quest and objectification, thus closely reflects an instrumental approach to knowledge, reminiscent of Bacon's notorious coupling of scientific understanding with the conquest of inanimate nature, often in the language of sexual mastery. Captivated by such a heroic quest, no wonder that Victor believes himself to be invincible: 'Life and death appeared to me ideal bounds, which I should first break through, and pour a torrent of light into our dark world. A new species would bless me as its creator and source; many happy and excellent natures would owe their being to me' (Shelley, 1998: 36).

Victor's zeal is already verging on obsession and the single-minded and dysfunctional nature of his pursuit prefigures his downfall. In the grip of an

---

[5] In Branagh's film version, Victor is ridiculed by a professor of anatomy for his outmoded views.

'unnatural stimulus' (1998: 36) the changing seasons pass him by. He abjures contact with family and friends. 'In a solitary chamber, or rather cell, at the top of the house, and separated from all the other apartments . . . I kept my work-shop of filthy creation' (1998: 36). By the time his researches near their conclusion, Victor is in a state of near-collapse: '. . . I became nervous to a most painful degree . . . I shunned my fellow creatures *as if I had been guilty of a crime*. Sometimes I grew alarmed at the wreck I perceived that I had become . . .' (1998: 38, my emphasis). The insistence upon the objectivity of scientific knowledge, that research itself is value free, is refuted by Mary Shelley's portrayal of Victor's *anomie* throughout the creative process. Slowly but surely, consumed by his mono-mania, Victor begins to lose his humanity. The moment at which Victor animates the being reveals him definitively as the modern Prometheus. Like his ancient predecessor, the 'thief of fire', Victor steps over the boundary between taxo-nomy and manipulation by summoning up the 'spark of being' (Mellor, 1995: 121), and unleashes the dangerous energy by which life itself is generated.

Victor's isolation is also signified by his gradual displacment from the orderly topography of his own native status. Ignoring family, he takes up residence in Ingolstadt, then flees to northern Europe towards the Arctic wilderness as he vies, like cat and mouse, with his misbegotten creation. His departure from the security of his roots becomes a metaphor for his trespass into dangerous know-ledge: 'Learn from me, if not by my precepts, at least by my example, how dangerous is the acquirement [sic] of knowledge and how much happier that man is who believes his native town to be the world, than he who aspires to become greater than his nature will allow' (Shelley, 1998: 35).

At the heart of Mary Shelley's analysis of the nature of 'Frankensteinian knowledge' is therefore the conviction that monstrosity is a result not of some inherent transgression against divine proscription, but a consequence of obses-sive, alienated pursuit of certainty and controlling power:

> The desire to control nature through science . . . is part of a larger desire for control and mastery . . . Mary Shelley . . . was concerned with this sort of ruthless mastery – the sort exhibited not only by fanatical scientists, but also by all-powerful rulers toward the native inhabitants of colonized territories. (Lesser, 1992: xv)

Whose monstrosity? Whose humanity?

Victor's ambition to create and renew life leads only to death. (Heffernan, 1997: 141)

Frankenstein may be read as an exploration of the power of emerging scientific practice both to enhance the quality of life but also, potentially, to disturb the delicate balance between life and death. This goes deeper than the solitary conduct of Victor's research. Themes of birth, life, death – natality and necrophilia – run throughout and form further fault-lines between humanity and monstrosity.

77

As Victor's incursions into the mysteries of science proceed, his methods are clearly informed by one key principle: the interconnection of life and death. 'To examine the causes of life, we must first have recourse to death,' he explains (Shelley, 1998: 33); and so his search for the sources of life leads him to the sites of death. 'Now I was led to examine the cause and progress of this decay and forced to spend days and nights in vaults and charnel-houses' (1998: 34). There, he is struck by the rapidity with which the bloom of life is transformed into decomposition. It is the tenuousness of the veil between death and life that first puts Victor on to the trail of his great scientific discovery:

I saw how the fine form of man was degraded and wasted; I beheld the corruption of death succeed to the blooming cheek of life; I saw how the worm inherited the powers of the eye and brain. I paused, examining and analysing all the minutiae of causation, as exemplified in the change from life to death, and death to life, until from the midst of this darkness a sudden light broke in upon me . . . After days and nights of incredible labour and fatigue, I succeeded in discovering the cause of generation and life; nay, more, I became myself capable of bestowing animation upon lifeless matter. (Shelley, 1998: 34)

Victor's obsessive quest for a scientific method that can cheat death is ostensibly motivated by an ambition to defy mortality; yet other elements of his behaviour suggest that more powerful than this is his profound ambivalence towards anything associated with birth. His fascination with mortality is matched by a disgust for all that represents the maternal and the procreative. Victor's upbringing is shadowed by the death of his mother, which occurs in the novel as a result of her nursing Elizabeth back to life from scarlet fever. More evocatively, perhaps, the Branagh film portrays Victor's mother dying in childbirth[6] – the baby in question being William, the boy who would, in turn, die at the hands of Victor's creation. The desire of Branagh's Victor to study medicine (and rapidly, by association, to restore the dead to life) is explicitly motivated by his mother's needless death.

Thus the creature's monstrous birth origins have their parallels in Victor's own primal desires. His monstrous impulses are forged amid the heated interplay of desire, mortality and the feminine. This helps to explain Victor's oscillation between abhorrence and attraction for his creature and his horrified embrace of the female body, both literally in the shape of Elizabeth and figuratively in his mother's. His flight from the creature when first he beholds it (thus setting a pattern of flight and desertion) culminates in a dream in which he encounters Elizabeth in the streets of Ingolstad. As he embraces her, she turns into a corpse 'livid with the hue of death' (51) before mutating into a nightmarish vision of his dead mother. A reunification with a loved one, whose beauty and vitality so starkly contrasts with the hideous wretch from whom he flees, dissolves into a

[6] Mary Shelley's mother Mary Wollstonecraft died from post-partum complications on 10 September 1797, ten days after giving birth to her daughter.

shocking reminder of the rotten core of death at the heart of his act of creation; that 'the miserable monster whom I had created' was also 'the demoniacal corpse to which I had so miserably given life' (Shelley, 1998: 40).

Victor's ambivalence towards the monster may be a reflection of the psychic turmoil he exhibits towards women. As both the site of birth and also the route into mortality, the maternal is viewed with wonder and horror – the transgression of the uneasy boundary between life and death. Like the monster's aw(e)fulness, the maternal body straddles ontological boundaries, representing a mixture of attraction and repulsion characteristic of the object of primal desire that is also forbidden. The encounter with the supposed object of his desire, Elizabeth, and the symbolic embrace of familial responsibility (both to honour the promise made at his mother's deathbed concerning his betrothal to Elizabeth, and the anticipation of his duty to continue the family name), triggers a deeper, more primal desire for his dead mother in a reiteration of the symbiosis of death and life.[7] This is further reinforced at the point of Elizabeth's death at the hands of the creature in 'virtually a parody of wedding-night sexual ecstacy' (Lesser, 1992: ix).

The self-contradictory desire to evade death that leads into the embrace of death, and Victor's inability to form relationships with the living, suggest a strategy of denial. He has experienced the interconnectedness of life and death, both in his mother's fatal sickness and in his own experiments, but it represents a primal confusion he seeks to resolve by repressing the realities of death. His method of coming to terms with this is an obsession with fabricated and artificial creation in the name of achieving mastery over the fragility of life. However, his successful animation of life out of death only serves to rekindle the confusion. This morbid evasion of life – symbolized so frequently by all that is associated with the feminine and characteristic of much of the Baconian world-view – is at the root of Victor's monstrous researches. I now turn, therefore, to the final area of the making of monstrous knowledge: the contest between natality and necrophilia within the novel.

NATALITY AND NECROPHILIA

What if, out of a deconstructive reading of the western tradition's preoccupations with mortality, we began to privilege natality, its repressed other? (Jantzen, 1998: 129)

---

[7] Victor's behaviour is, in psychoanalytic terms, classically narcissistic. He reminisces that his happiest memories are of days of youthful innocence when he was not expected to exercise an adult role. Accountability for his actions, adult sexuality and paternal ties to his creature are beyond him. Victor is inclined to fall away into a dream or a hysterical swoon at the first sign of a crisis: when the creature first comes to life, and on his wedding night when he remembers the monster's terrible threat against him. These may be hysterical devices to evade responsibility at times of stress, although such little deaths (swoons and seizures) may also signal Victor's morbid necrophilia.

According to Grace Jantzen, the Western philosophical tradition is obsessed with death. She sees this trait exemplified in the model of Baconian science in which human dominion over nature, flight from contingency and the denial of embodied, gendered and situated subjectivity dominate the epistemological paradigm. Paradoxically, therefore, life becomes a constant struggle to deny those aspects of human experience that remind us of our mortality. Western modernity thus exemplifies the same kind of mixture of repulsion and attraction with death that, in Chapter two, I characterized as typical of reactions to the monstrous. For Jantzen, the Western symbolic system has developed a 'necrophilic' imagination, precisely to address the ambivalence felt in the face of mortality, corporeality and contingency. In order to expel the monstrousness of life and risk, it is necessary to privilege that which supposedly 'conquers' death: 'If humans are to find meaning in a life which moves inexorably towards death, one strategy for dealing with anxiety is to postulate immortality and a God who guarantees it, especially . . . if that God also authorizes mastery over that which reminds of mortality: women, bodiliness, and the earth to which we all return' (Jantzen, 1998: 131).

The necrophilic imagination is apparent in such qualities as 'a drive to infinity: an insatiable desire for knowledge, a quest for ever increasing mastery, a refusal to accept boundaries' (Jantzen, 1998: 154). It is also reflected in the militarism of the foreign policies of Western states, in the violence of popular entertainment and in the economic priorities of the West. It can be seen in the sale of armaments to the Two-Thirds World in preference to the funding of basic aspects of subsistence such as clean water and primary health care.[8] Similarly, fascination with 'other worlds', such as extraterrestrials and UFOs, and religious discourse about 'eternal life after death' do not represent a healthy interest in the flourishing of life, but a pathological denial of the immanent and the material. 'Finitude is not evil. Rather, it is the effort to conquer finitude instead of treating it with respect which has been the cause of much evil, much suffering' (Jantzen, 1998: 155).

Such an obsession with immortality does violence to the body and distorts gender relations and representations. Insofar as the prospect of mortality and physical contingency endangers the quest for immortality and transcendence, the quest for true wisdom is compromised – one might say mortified – by corporeality. In order to concentrate on the life of the spirit, one must therefore learn to master the passions of the body. Women, associated with matter and the body, become repositories of mortality – a rationalization for their subordination.

---

[8] See Chapter seven for further discussion of the way in which the attraction of advanced technologies may obscure critical analysis of the priorities of global corporate interests in determining future economic development, at the expense of the poorest of the poor.

However, the centrality of death within the Western symbolic rests, un-acknowledged, on the repression of an alternative symbolic, that of 'natality'. This is not, as Jantzen hastens to say, merely a valorization of motherhood and childbirth as women's destiny, nor is it an attempt to deny human mortality. It is rather a recognition that humanity has a shared origin in birth which neces-sarily embeds us in common experiences, both biological and social, and com-mits all living beings to sociability, interdependence and embodiment. After Hannah Arendt, Jantzen argues that the reality of birth, not resistance to the prospect of death, guarantees our freedom and authenticity. To return to Victor, in his scientific pursuits, and especially in his self-imposed isolation and indif-ference to the seasons, he demonstrates what Arendt terms 'worldlessness' (Jantzen, 1998: 151), exhibiting a separation from his fellow humans and the environment. 'I pursued nature to her hiding-places', he says; ' . . . I seemed to have lost all soul or sensation but for this one pursuit . . .' (Shelley, 1998: 36, my emphasis).

His self-absorption shows some of the traits of necrophilia. His forays into the charnel-houses betray a pathological fascination with death, even as he seeks to defy it through his experiments. His drivenness towards his scientific goal is designed to cheat death and to create a new race of immortals. Meanwhile, however, in his dreams women, carnality and death are intertwined in mon-strous couplings (Shelley, 1998: 39).

The 'unedifying spectacle' (1998: 141) of Victor, like other philosophers before him,[9] seeking to give birth without women shows a vain quest for pro-creation without embodiment – a desire to supplant nature and confound natality. A model of scientific knowing expressed in the language of subordina-tion and possession, disclaiming any affinity with non-human nature and alien-ated from any sense of moral responsibility is, Mary Shelley suggests, likely to engender monstrous results. This portrayal of Victor's 'typically insensitive and self-absorbed' (Butler, 1998: xl) pursuit of nature typifies what Marilyn Butler calls 'civilized degeneracy' (Butler, 1998: xli). Arguably, therefore, Victor's neglect of his family, his creature and his civil responsibilities is the source and cause of the monstrosity – not his pursuit of knowledge per se but a denial of the very affinities and connections that constitute his humanity, in the shape of his lack of compassion towards his own creation, his abandonment of his family and a careless disregard for the nature he so wilfully plunders and manipulates. In exposing Victor's necrophilic methods as the reason for his inability to bring forth a truly natal creature, we are brought face to face with the fullness of his – not the creature's – 'natal alienation'. The question of what might constitute illegitimate science is not about halting at forbidden knowledge (as theological

[9] Plato, through Diotima, declares that the best kind of procreation does not involve women, but takes place in the meeting of (male) minds; Francis Bacon; Kant's view that God 'fathered' the world, thus perpetuating a profound dualism between nature and reason (Jantzen, 1998: 142).

prohibition), nor about the consequences of *dangerous* knowledge unleashed (a version of consequentialism – intervention in the processes of nature serve to unbalance its equilibrium), but becomes a question of *irresponsible* knowing, a disregard for the affinities and connections in the scientific process (Allen, 1992: 307–9). 'Uninhibited scientific and technological development, without a sense of moral responsibility for *either the processes or products* of these new modes of production, could easily, as in Frankenstein's case, produce monsters' (Mellor, 1995: 134, my emphasis).

Yet Jantzen's thesis causes problems for Victor's unfortunate creature. Unborn, made from fragments snatched from death, a singleton of his species, the creature is marooned on the wrong side of that human sociability so warmly commended by Jantzen when she states, 'Birth is the basis of every person's existence, which by that very fact is always already material, embodied, gendered and connected with other human beings and with human history' (1998: 141). A literalist reading of this would exclude Victor's creature from the comforts and privileges of human community, regardless of his moral or intellectual standing. By implication, however, would it also disqualify any hypothetical form of post/human or artificial life not organically 'born'? Is such a construct of natality merely colluding with Victor's desire once more to banish the creature beyond the pale of acceptable humanity? It is not the intention of this book to adjudicate on whether genetically modified or cloned humans or forms of artificial intelligence might be considered 'human'; but I have been concerned to consider how the various representations and attributions of monstrosity in Frankenstein offer models of what it means to be truly human, and there is no reason not to speculate about the ways they might be applied to possible forms of post/human lives. It is surely not the literal fact of biological birth which Arendt's and Jantzen's ethic of natality is commending, but rather the values people might embrace throughout their entire life-cycle. If so, then an individual who was not a biologically natal human might still be considered worthy of the attributes of personhood. Indeed, Mary Shelley invites her readers to contemplate whether humanity – in a moral rather than an ontological sense – is assumed at birth or acquired through socialization: 'Mary Shelley saw the creature as potentially monstrous, but she never suggested that he was other than fully human' (Mellor, 1988: 63).

For while the creature possessed no personal name in the original novella, Mary Shelley's experiments with radical literary methods afforded him a voice in the context of a narrative of multiple and competing authors. This makes it all the more difficult to know definitively who and what is authoritative and normative, and who is intended to excite readers' fear, support, approval and compassion. The reader leaves the arctic wastes, the site of Frankenstein's final confrontation with his creation, undecided as to many endings and resolutions.

There are many representations of what it means to be human in *Frankenstein*, refracted through the attributions of monstrosity that ultimately implicate Victor as well as his creation. In classical teratology, the birth of a monster or prodigy signals a rupture in the taxonomic integrity of human nature. Certainly Victor's interventions have violated the boundaries between death and life, artificial and natural, made and born; and as the object of his intentions, the creature breaches the rules of natality as a condition of what it means to be human(e). Yet in other ways, the creature's own demeanour requires his society – and perhaps the readers themselves – to re-examine the very criteria by which they think about what it means to be human. It is impossible to have spent time in the creature's company not to feel some generosity towards him, and the text invites the reader to question the ontological hygiene that might forbid Victor's creature even an adoptive humanity. An ethic of 'natality' helps to expose Victor's dysfunctional and necrophilic impulses; but it should also assist in expanding predominant understandings of what it means to be human, a process which the creature's own testimony seems to demand.

CHAPTER FOUR

Body of clay, body of glass

What the Golem legend can teach us is that the Golem, the machine, while not human, is nevertheless a reflection of the best and the worst of that which makes us human. (Sherwin, 1985: 48)

Over the past quarter-century a significant subgenre has emerged in science fiction of work which turns the traditions of utopian and progressive optimism towards an examination of gender relations. Early examples such as Ursula le Guin's *The Left Hand of Darkness* (1969), Joanna Russ's *The Female Man* (1972), Samuel Delany's *Triton* (1976) and Sally Gearhart's *The Wanderground* (1979) are generally regarded as the most distinguished contributions of this time (Russ, 1995: 133–48), using the 'fabulation' of alternative futures and alien cultures to imagine new configurations of gender roles, identities and relationships. A later example, Marge Piercy's *Body of Glass*, published in 1992[1] stands in this same tradition, using futuristic and technologically sophisticated settings to explore a number of themes to do with culture and gender. *Body of Glass* is set in 2059; the heroine, Shira, escapes from the benevolent but smothering regime of Yakamura-Stichen, her multinational corporate enclave, to return to Tikva, the city of her fore-mothers.[2] This utopian society – inspired by the kibbutz movement of twentieth-century Israel – harnesses genetic, digital and cybernetic technologies in pursuit of egalitarian values.

There are, however, many levels to *Body of Glass*. Drawing upon the disparate conventions of 'cyberpunk' fiction (see also Chapter eight), the cyborg writing of Donna Haraway as well as feminist utopian writing, Piercy has written a novel bursting with post/human lives. All the characters have been artificially altered in some way. Assisted reproduction is commonplace, and women rarely carry children to full term *in utero*. Bodies and minds are modified by prosthetic

---

[1] Published in the US in 1991 under the title *He, She and It*.

[2] One of the first Zionist settlements in Israel was known as Petah Tikvah, or 'opening of hope'.

devices, pharmaceutical implants or by genetic engineering. Tikva's environment must be carefully controlled to resist contamination from pollution and nuclear fall-out. At the centre of this community is a 'cyborg' called Yod.[3] As a machine programmed to be almost-human, Yod may not be a monster, but he stands at Jeffrey Cohen's 'gates of difference' in his divided loyalties between programming and learning, duty and love, distinctiveness and assimilation. Piercy interrogates the basis on which the reader might attribute humanity to Yod – or, indeed, any of the characters; and whether a digital and biotechnological age will develop into a dehumanized dystopia or present opportunities to transform 'human nature' for the better. Piercy is especially concerned with the possibilities to re-envision things in particular directions: in terms of more egalitarian relationships between women and men, a greater integration of humanity and the environment, and less alienation between humanity and its artefacts.

A further level still to *Body of Glass* is Piercy's use of parallel narratives, a device frequently used to depict alternative universes, present in Russ's *Female Man* and, indeed, in Piercy's earlier novel *Woman on the Edge of Time* (1976). In the latter, Connie, the heroine, shifts between three parallel dimensions: a psychiatric ward in her own time, a polluted, socially divided society of the future and a pacific, gender-inclusive community of the same period. In *Body of Glass*, Yod's mentor, Malkah – who is also Shira's grandmother – gives him a retelling of the Jewish legend of the Golem of Prague. This is a story which emerged from the Jewish communities of Central Europe during the early modern period and gained wider popularity in the mid-nineteenth century. It concerns the creation of a magical creature conjured from clay, 'a man created by magical art' (Scholem, 1965: 159) who defended the Jewish community of Prague against persecution over a century earlier. Like the golem of the legend (whom Piercy calls Joseph), Yod is a product of human artifice; like Joseph he has been created to defend his people against an external threat.[4]

What is a work of feminist science fiction doing in appropriating this tale? Why does a futuristic story of cyborgs, robots and androids – also featuring virtual reality, a digital Net and information hackers – introduce an ancient Jewish legend? One answer is that Piercy has frequently drawn on her own Jewish heritage in much of her poetry and other writing (Piercy, 1995). Similarly, she is not the first to have her creative imagination fuelled by the golem; there is speculation that Mary Shelley knew of the legend. The founder of modern

---

[3] Yod is actually an android, not a cyborg, as many commentators have noted (Fitting, 1994: 5). Piercy herself acknowledges the influence of Donna Haraway's 'Cyborg Manifesto' on *Body of Glass* (Piercy, 1992: 583–4), which may explain the terminology. Yod is the first letter of the Tetragrammaton, the letters denoting the sacred name of the Holy One: Y, H, W and H. In the Kabbalah, Yod was seen as the fundamental revelation of the Creator God.

[4] In Yod's case, it is the incursions of the multinational corporation that once employed Shira, Yakamura-Stichen, which is anxious to undermine the independence of Tikva.

cybernetics, Norbert Wiener, places the clay golem in a continuum of artificial beings that stretches from antiquity to twentieth-century information systems (Wiener, 1964);[5] and it is possible to identify affinities between the golem and other fantastic creatures of the Western imagination, such as talking statues, homunculi, tragic monsters and even robots. In fact, the golem has a rich and varied past – a history that in its many layers of telling and retelling might best be thought of as a *genealogy*, comprising an identity which betrays multiple origins and discontinuities rather than a seamless and uninterrupted line of descent.

In the shifting narratives of the golem legend, we can see how one single telling of a myth is insufficient. It is a many-textured story, developing over time. As a creature on the boundaries of the human and almost-human, the golem is not one but many representations of different preoccupations with what it means to be human through various epochs. Just as Foucault argued that no single humanist identity continues unbroken, so in the different manifestations of the golem there is a pluralism of definitive narratives, about cosmology, about invention and about identity – Jewish, gendered and post/human. As with Frankenstein's creature, the golem's shifting identity shows forth the fault-lines between the things culture has chosen to call persons and those it calls machines. With each retelling, these boundaries are constantly being reconstituted, and their shifting contours reveal as much about the interests, duties and preoccupations of the originators of the stories as they do about any real essence of the golem. As a creature stationed at the gates of difference, the golem is therefore an index of the changing paradigms of normative and exemplary humanity. It is his status as fabricated being[6] and the way in which his changing identity functions as a foil to human enquiry that is of interest. As attributes of the golem are re-moulded and reshaped over the centuries, it is possible to see them as vehicles for human reflections on the origins of the cosmos, on human liberty to emulate divine creativity and the problematic character of post/human identity in a technoscientific age.

What strikes the student of golemry, therefore, is the creature's (and the legend's) chameleon-like nature. The enduring appeal of the golem rests on his simultaneous capacity to absorb a myriad of cultural influences while serving as a consistent vehicle for perennial concerns. In weaving the golem into her novel, therefore, Piercy is honouring a long tradition of storytelling. The 'Body of Glass' of the British title may be intended to evoke test tubes or the glass retort of Paracelsus' homunculus (see later), but it serves as a helpful metaphor in other ways. Like the glass shards of a kaleidoscope, the many fragments of the golem story come together to create many patterns and pictures as each twist of

---

[5] The first supercomputer at the Weizmann Institute in Rehovot, Israel, was named Golem.

[6] Most tales feature the golem as male, with the exception of a seventeenth-century version concerning one R. Shelomo ben Gabirol, who created a female housekeeper out of wood (Idel, 1990: 233).

the prism creates a new design. And like a looking-glass, the golem also reflects an image back to us, of representations of the human refracted in the shape of the almost-human.

On the golem and its symbolism

We need to go back to the beginning, to the earliest roots of the idea of the golem. Talmudic[7] commentaries on Psalm 139 characterize the psalm as a hymn of praise from Adam to God. The psalmist speaks of the formation of human life from the mother's womb, and also from the depths of the earth:

(v. 13)   It was you who created my conscience;[8]
           You fashioned me in my mother's womb.
14   I praise you,
           For I am awesomely, wondrously made;
           Your work is wonderful;
           I know it very well.
15   My frame was not concealed from You
           when I was shaped in a hidden place,
           knit together in the recesses of the earth.
16   Your eye saw my unformed limbs [golmi];
           they were recorded in your book;
           in due time they were formed,
           to the very last one of them.
           (Psalm 139, vv. 13–16. Jewish Publication Society, 1985: 1274)

The term golmi/golem (v. 16) comes from the root galam, meaning to enfold, or wrap, which is frequently translated as 'embryo': the not yet formed and still folded (or enclosed and cocooned) human body. As well as alluding to the embryo or unformed being, 'golem' can also suggest a malformed or deficient person, such as the unmarried woman or man, who is incomplete without a partner and who has yet to fulfil their potential in terms of the continuation of the human race (Idel, 1990: 232).

It is somewhat misguided, therefore, to regard this passage as simply the precursor of later legends about an artificial creation (Schäfer, 1995: 254). Psalm 139 is essentially about the glorification of God in and through creation, and of the creatureliness of humanity. In the etymological link between Adam ('adamah) and the earth (v. 15), the Psalmist is alluding to the tellurian nature of humanity. The genesis of the earth as a weaving together, or 'knitting', is thus of a piece with the creation of humanity itself. It is not too fanciful to see the symbolism here of the golem as a thing which testifies to the wonders of creation, perhaps also prefiguring the covenant between God and Israel that

[7] Rabbinical commentaries deriving from the fifth and sixth centuries CE.
[8] Literally, 'Kidneys'.

was believed to be present, albeit in unfinished terms, from the very beginnings of time.[9]

Legends about human *emulation* of divine cosmogony begin to emerge from about the third to the fourth centuries CE. This extract, found in the Babylonian Talmud, constitutes the textual heart of the golem myth:

> Rava said: If the righteous wished, they could create a world, for it is said: 'But your iniquities have separated you from your God' (Isa. 59.2). Rava created a man and sent him to R. Zera . . . [Zera] spoke to him but he did not answer. Thereupon he (the rabbi) said to him (the artificial man): You are coming from the fellows – return to your dust!
>
> R. Hanina and R. Oshaya spent every Sabbath eve in studying the instructions concerning creation (*hilkhot yezirah*), and a calf one third of the natural size was created by them, and they ate it. (bSanh. 65b, quoted in Schäfer, 1995: 252–3)

These two stories are intended to mirror each other. They are essentially reflections on the nature of human relationship to the divine, and not a technical manual. As Peter Schäfer points out, 'We do not have much of an artificial man here . . . and we do not have much magic either . . .' (Schäfer, 1995: 254). Moshe Idel suggests that these texts were polemics against idolatry, and especially against pagan traditions of animated statues (Idel, 1988: 19). The mystics, or learned ones, in these stories, who sought to study the mysteries of creation and to emulate them, still fall short of God's perfect feats, despite their piety. The creation of a speaking (therefore rational and therefore perfect) being in the divine image would be a test of their moral perfection; but even the powers of the most pious are limited by their iniquities. These stories may rather be viewed as ironic tales of outstanding failure, attesting to the flawed powers of humanity. The identity of the silent golem is rapidly unmasked, underlining that the moral of the story derives from his defective status, rather than the circumstances of his creation. Even if Hanina and Oshaya are to be considered idolators, they only succeed in bungling their transgression, yielding no more than a small meal.

Although Emily Bilski and Moshe Idel regard the Talmudic story of Rava and Zera as offering definitive proof 'that Judaism accepted the idea of the creation

---

[9] Later rabbinic interpretation continued to reflect on the significance of this in the creation of Adam in the Genesis stories. Talmudic commentaries on Genesis stress the etymological link between Adam and the earth: Adam is the one made from the clay of the earth. Further commentaries suggest that Adam remains a golem until the rest of creation is finished, to avoid any suggestion that he was a co-creator with God (Scholem, 1965: 162). Only once the whole of the universe is in place is the protohuman granted speech and reason, deemed an essential aspect of being fully human. Another tradition teaches that when Adam was first made as a golem he filled the entire earth, thus enabling him to fulfil the words of Psalm 139, of God's being able to see him. An elaboration of this speaks of the future history of Israel being revealed to Adam: thus, the unformed protohuman is shown the destiny of an as yet unformed nation (Schäfer, 1995: 251), suggesting the privileges of a covenantal relationship present already in embryonic form, just as the (Jewish) human race awaits completion.

of an artificial man' (Bilksi and Idel, 1988: 10), this text is primarily part of a continuing speculation on the relationship between divine creation and human demiurgy. The artificial creatures are minor players, inserted to illustrate a theological argument. The perilousness of human creativity and the ironic nature of human emulation of the divine to which these tales allude are still different from stories of hubris, which reveal the inevitable dangers and tragedies of human meddling in creation. While created in the likeness of God – possessing reason, speech and creative agency – humanity can never fully achieve the creation of life.

This is also part of a contiguous strand of Talmudic teaching which drew a direct connection between language and the cosmos, teaching that God first read the Torah then created the world (Alexander, 1992: 238). Words and the utterance of speech were understood to lie at the very heart of divine cosmogony; and by extension, human language was held to constitute the emulation of divine qualities. The animation of the golem via evocation of letters is merely the activation of a pre-existent divine indwelling (Idel, 1990: 248–9), consistent with a proto-Gnostic view that a hidden Supreme power suffuses the universe, awaiting activation by human demiurgy.

SEFER YETZIRAH

Belief in the animating power of language represents an important continuity between Talmudic tradition and that of the *Sefer Yetzirah*. *Sefer Yetzirah* is a major text of Jewish spiritual practice, thought to date from the fourth century CE. The central concern of the *Sefer Yetzirah* is cosmological, identifying '32 secret paths of Wisdom' which comprise twenty-two letters of the Hebrew alphabet plus ten sacred *sefirot*, or dimensions of divine utterance.[10] The implication is that by combining letters God makes both created life (*yezur*) and speech (*dibbur*). *Sefer Yetzirah* propounds a link between Hebrew letters, parts of the body (especially limbs) and the arrangement of the stars and planets into constellations (reflecting the influence of pagan astrology). The processes of creation are enacted through linguistic combination, for hidden within every living thing is a secret configuration that reflects the divine power emanating through the entire cosmos. So the combination of all twenty-two letters of the alphabet forms both a sacred name – 'the unified crystallization of the alphabet' (Idel, 1990: 11) – and a living body or anthropomorphic structure.

Sefer Yetzirah is both a treatise on the origin of the universe and implicitly a model lesson in the pursuit of the true path of faith by emulating divine creative

---

[10] The universe was held to have been created by the combination of all twenty-two letters of the alphabet fixed on a wheel or wheels. There are 231 combinations of the letters, divided into gates, which move backwards and forwards. As the wheels turn, syllables are formed. In some traditions, the letters are combined with the Tetragrammaton (Idel, 1990: 10, see n. 3, p. 85).

activity (Idel, 1990: 15). 'Every process in the world is a linguistic one, and the existence of every single thing depends on the combination of letters that lies hidden within it' (Scholem, 1971: 508).

The concept of the *sefirot* as material signs of the hidden wisdom of the divine – 'the numerical infrastructure of the cosmos' (Idel, 1990: 9) – reflects a Gnostic or Neoplatonic strain within the *Sefer Yetzirah*, which probably occurred under the influence of the Pythagorean revival during the fourth century CE.[11] Many of the principles of the *Sefer Yetzirah* are modelled on a Pythagorean understanding of the universe as a 'metaphysical dance of numbers' (Wertheim, 1997: 25). These primordial numbers which constitute the essence of reality can also be transposed with letters, as emanations of esoteric but redemptive knowledge. Like the later Gnostics of the early Christian era, therefore, ancient Pythagoreans held that the material realm is ruled by a malevolent simulacrum of the true god. If not the dualism of a supreme divinity obscured by the acts of an evil demiurge, then this view certainly understood that the hidden mysteries of creation were most tangibly accessible to humanity through the material manifestations of symbols. Authentic knowledge of this divinity – true *gnosis* – only illuminates the human spirit via the emanation of sacred wisdom in the form of 'sacramental procedures, secret names, and magic formulas that enable the soul to break through the vagaries of nature under demiurgic control and mount to God' (Davis, 1993: 604). This *gnosis* is manifested through particular channels, such as numbers, letters, points of the compass, even parts of the body. All are sacred by virtue of their being the material manifestations of an unblemished world, the embodiment of a heavenly reality that is essentially beyond words.

KABBALAH AND GOLEM

The emergence of the movement known as the Kabbalah[12] from the twelfth century CE gave the *Sefer Yetzirah* a renewed prominence. In assimilating earlier traditions of Gnosticism and Neoplatonism, the practitioners of the Kabbalah stressed the esoteric nature of religious experience whereby knowledge cannot be apprehended directly but only alluded to in symbolic form (Scholem, 1965; Fine, 1984). If religion consisted of a system of mystical symbols that pointed to the hidden mystery of the universe, then language assumed a particular prominence as embodying the animating power of creation and serving as the means of communication between the Holy One and humanity. But as well as reading it as a meditation on cosmogony, kabbalistic interpretation of the *Sefer Yetzirah* yields up a more magical and practical reading: a belief in the power of incantations – combinations of letters and words – to manipulate the created order.

---

[11] Born in the sixth century BCE, Pythagoras travelled widely, including to Persia and Egypt; he is held to have introduced mathematics to Greece from Egypt.

[12] Literally 'tradition', and not well translated as 'mysticism' because of its strong roots in collective scholarly pursuits as much as its emphasis on personal interiority or spirituality.

This, combined with interest in esoteric arts and ritual, encouraged leading figures to produce texts elaborating on the *Sefer Yetzirah* which set out rituals for the animation of a clay figure using combinations of sacred letters.[13] Yet the practitioners of the Kabbalah would have regarded such practices as part of the spiritual discipline, practical rituals that, by metaphorically participating in divine cosmogony, would engender states of mystical ecstacy. It was not a profaning or sacrilegious act to invoke the sacred art of letters, because the intention was to mirror the divine creative act, rather than to engage in magical manipulation of the material world, or to dabble in demonic powers, or even to displace the Holy One. Indeed, one strand of neo-Gnostic Kabbalah taught that creation was essentially unfinished until humanity participated in demiurgy (Scholem, 1965: 174).

The greater emphasis in commentaries on the *Sefer Yetzirah* on the fabrication of a clay anthropoid is most likely due to the specific influence of its Kabbalistic interpreters. Central to the Kabbalah is an understanding that the human body is in some way the manifestation of the secrets of the heavens. The early Kabbalists believed that the infinite dimension of God (*Ein Sof*) was unfathomable and unattainable. However, the concealed qualities of God emanated creation through the ten sacred *sefirot*: a kind of divine radiance or illumination (Fine, 1984: 318). Each *sefirah* symbolized specific aspects of the material world, or divine virtues, or colours; but also corresponded to a part of the body – a limb, sexual organ – or emotional and intellectual faculties. One of the terms for the entire *sefirotic* system is 'Adam Kadmon', or primal being (Fine, 1984: 324). Thus, within the Kabbalistic cosmology, there is a clear link between the human form and the sacred emanations of divine mystery: '[O]ne of the ways in which the *Sefirot* are imagined is in the guise of the human form. Thus our bodies reflect and therefore symbolize the constituent elements of the life of God' (Fine, 1984: 326).

To the followers of the Kabbalah, therefore, not only were human beings made in the image of God by virtue of rationality and creativity, but embodied the totality of the *sefirotic* system. Humanity's physical form was suffused with

[13] These techniques seem to have been current among the German Hasidim and French and Spanish Kabbalists, and one commentary, that of Eleazar of Worms (c. 1165–1230) on the *Sefer Yetzirah* has been dated to the end of the twelfth century CE; that of the Spanish spiritual master Abraham Abulafia (1240–92) about a century later. There was significant variation between different Kabbalistic schools (Scholem, 1965: 184–91). In some versions, dust is gathered from the four points of the compass and moulded into an anthropomorphic shape, followed by the recitation of permutations of letters. Recitation of the letters backwards would unmake the golem. Other rituals involved placing a clay figure in a hole in the earth, further evoking connections between burial and rebirth and the golem's tellurian origins (Idel, 1988: 22). Another variation entailed circumambulating the clay figure in a ritual circle dance, mimicking the circular gates of letter combinations in the *Sefer Yetzirah* (Pseudo-Sa'adiyah, thirteenth century: Idel, 1988: 20). At about the same time a tradition emerges of the golem bearing the letters for 'emet (truth) on its forehead. The unmaking of the golem occurs when its maker strikes out the 'aleph, leaving met (death).

divine energy. Rituals animating a clay figure therefore drew upon the understanding that the syllables uttered in creation correspond to parts of the body. By creating an anthropoid the spiritual masters were symbolizing a reminder or reflection of the sacred speech at the heart of creation (Scholem, 1971: 608). The parallels between certain permutations of the alphabet and parts of the human body should be understood as an extended analogy between the body and the cosmos – that the human body is a microcosm of the universe – rather than a set of instructions on the creation of an animated being:

The procedure seems to be self-sufficient, a performative magical act the purpose of which is to imitate God, certainly not to make any use of the golem as is the case in the much later sources. It is only the ritual that matters: by using Sefer Yesirah. . . . man can imitate God's creative act. (Schäfer, 1995: 260–1)

The silence of the golem

The dominant tradition of the golem is of a paradoxically silent creature, created through words but proven imperfect by his non-participation in the world of words – and therefore excluded from partaking in the sacred knowledge held to underpin the very cosmos itself. However, a few apocryphal fables do allow the golem to speak. A French, probably Kabbalistic, text of the thirteenth century has the golem delivering a terrible warning, one of the first sources to hint at the ambivalent or dangerous potential of such a creature. The prophet Jeremiah, after the study of the Sefer Yetzirah, creates a golem in the traditional fashion of 'combination, grouping and word formation' (Scholem, 1965: 180). The inscription on this particular golem's brow is YHWH Elohim 'emeth (The Lord God is truth). The golem laments his creation, and scratches out the 'aleph, making the text read, 'The Lord God is dead.' When questioned, the golem delivers a parable from which he concludes that once humanity possesses the ultimate powers of creation, people would have no need of God. Jeremiah decides to destroy the golem by reversing the ritual, and the story concludes with his sorrowful evaluation, 'Indeed it is worthwhile to study these matters for the sake of knowing the power and dynamis [sic] of the creator of the world, but not in order to do [them]. You shall study them in order to comprehend and teach' (Idel, 1990: 67).

This story exemplifies debates about the wonders and dangers of demiurgy. By bringing forth life, humanity emulates such divine power; but such an act also carries with it the dangers of blasphemy, in that human wonder at such miraculous deeds might obscure their divine origin, thus bringing about the very effacement of God:

Golem-making is dangerous; like all major creation it endangers the life of the creator – the source of danger, however, is not the golem or the forces emanating from him, but the man himself. The danger is not that the golem, become autonomous, will develop overwhelming powers; it lies in the tension which the creative process arouses in the creator himself. (Scholem, 1965: 190–1)

The Talmudic tradition of a mute golem is further reinforcement of a world-view of creation coming about through the combination of letters. Yet there is still a certain paradox to this. Despite its linguistic origins, the golem exists outside the margins of speech. In the psalmists' depiction of Adam, bolstered by later midrash,[14] we find the association between the golem and human beings unfinished. Indeed, this perspective extends to the tale of Rava and Zera, where the golem is conspicuous by his muteness. At this level, the golem's exclusion from the world of words chiefly signifies his deficiency of powers of reason (Idel, 1990: 264). It also reinforces the idea of the golem as an instrument of demiurgic emulation, showing forth the imperfections of its creator, rather than a deliberate usurpation of divine power. The golem's silence places him outside language, exemplifying the futility of (fallen) humanity's efforts adequately to mirror the divine powers. It is not the golem's nature per se that interests the early tradition, so much as the discrepancy between divine and human attempts to create life.

But what constitutes this 'divine way of creating' and what is the connection between the golem's location beyond the bounds of speech and Jewish understandings of 'the Word'? As I have already indicated, the world-view within which the Talmudic commentaries and the deliberations of the Sefer Yetzirah took place regarded creation as brought forth via the power of language. Language in human usage had the potential to be creative in a similar way to that 'at the moment of creation when the world was generated by language' (Idel, 1990: 277). The Word, as the source of life, is the agent of the Holy One, and the text is sacred as the material representation of the divine intent. However, language never facilitates totality, mastery or absolute truth. Indeed, the Hebrew Bible teaches that God is beyond names. God may create through words but is ultimately not reducible to words. Creation itself is therefore characterized by 'metaphysics of absence, of unknowability, and of the unrepresentability of central truths' (Porush, 1998: 55). God suffuses all language, as language constitutes creation; but God is simultaneously present in the word and also unrepresentable in words.

---

[14] Midrashic literature comes in many forms: homiletic expositions on Scripture, exegesis of individual verses of the Bible, or stories and apocryphal legends on Biblical characters or eminent rabbis. Midrash refers not to a canon of texts, so much as an approach to the interpretation of Jewish scriptures and other texts, 'a kind of process or activity' (Holtz, 1984: 177) that was at its zenith between the fifth and thirteenth centuries CE. Midrash exists as a kind of literary supplement which comments, elaborates and debates points of the text. Such midrash, of course, then generates further midrashim.

93

If the golem is outside language, inhabiting the parergon[15] outside the text, then its silence only serves to remind us anew of the workings of language. Silence and speech never stand in isolation; because of *différance* even silence is implicated in language. *Différance* is not the 'ground of being' upon which language and meaning finally come to rest but the restless deferral which always places under scrutiny the very processes of calling into silence and speech: 'As soon as there is writing . . . there has always been *différance*, and it has immemorially promised that we are inscribed in language, even when we remain silent' (Hart, 1997: 163).

In the thirteenth-century text's elaboration of Jeremiah, cited earlier, where the golem *does* have speech (temporarily), it becomes clear that his continued existence constitutes a threat to the holiness of the people. The risk is not primarily that of idolatry – the creation of an anthropoid – but blasphemy, for the transgression comes in the existence of a being who might encourage mere mortals to believe that the ineffable qualities of the divine can be reduced to a magic formula. The creator of a golem tempts the witnesses of such a miraculous act not only to violate the commandment against graven idols, but also that prohibiting the objectification or displacement of the divine through any system of representation, pictorial or linguistic. The same world-view that informs the codification of such imperatives is already implicit in the qualities of language and text; and this emphasis on deferral, instability and open-endedness may thus be a theological statement in itself. Porush paraphrases parts of the Decalogue accordingly:

(Iconoclasm) Discard the idols, they are pictographic incarnations, much too pictographic for Me; (Ambiguity and Deferral) Do not take My Name in Vain. (Abstraction) I am the Unpronounceable God of Becoming . . . the abstract, portable, unknowable God-Who-Requires-Interpretation, and you still won't be able to fathom Me. (Porush, 1998: 56)

The problematic absent God is not approachable through naming or representation, but, in Edmond Jabès's terms, is glimpsed only in the 'exile' of silence, whiteness and blank space (Linafelt, 1997: 228–9). If God is represented by non-representation, never speaking directly but only manifested obliquely in the gaps of language, it may serve as a reminder that the full story of creation has yet to be written, and that human knowledge of the divine is deferred and incomplete. For the golem to live outside the linguistic realm, therefore, pronounces him, like YHWH, beyond the controlling signification of human culture. Rather than one who falls short, a misbegotten creature – like writing is to speech in Heidegger's philosophy, the monstrous simulacrum to the perfect creative act of will (Derrida, 1967: 3–5) – the golem may alternatively be rendered as defying containment and definition, embodying – showing forth, even – unattainable mystery.

[15] A term found in deconstruction, denoting the spaces and blanks between the lines and around the margins of text. A literal translation might be 'work subsidiary to one's employment' – a 'sideline' perhaps?

The servant

It is thus not until relatively late – probably the seventeenth century – that accounts of a living, viable golem start to enter the canon. The first such tale about the golem seems to have been that concerning the sixteenth-century German Hasidic master Eliahu of Chelm (d. 1583). This folktale is recorded in an anonymous, possibly Polish, manuscript, dated c. 1630 (Bilski and Idel, 1988: 13). Moshe Idel records the following extract from this earliest recorded version:

And I have heard, in a certain and explicit way, from several respectable persons that one man, [living] close to our time, in the holy community of Helm, whose name is R. Eliyahu . . . who made a creature out of matter [golem] and . . . it performed hard work for him, for a long period, and the name of 'emet was hanging upon his neck, until he finally removed for a certain reason, the name from his neck and it turned to dust. (Idel, 1990: 208)

The author is at pains to establish the veracity of the tale (it comes on good authority, despite the lack of hard evidence), although the nature of this manuscript is more one of folk legend than Kabbalistic commentary. It is likely that in this text we have the beginnings – what Idel calls the 'blueprint' (1988: 31) – of the later legend of the Golem of Prague. A creature is created by a specific historical figure, for a purpose; it grows, but is eventually destroyed.

What factors prompted the emergence of the viable, servant golem? The critical turning point in the evolution of the golem was probably the commentary by Eleazar of Worms in the thirteenth century CE. Before then, the inclusion of a clay figure is an imaginative (rather than experimental or material) device intended to symbolize the affinities between the human body and the cosmos. An ambulant golem who can be put to work only appears much later, and emerges due in part to the connections between Kabbalistic scholarship and the hermetic revival of the European Renaissance. This new conception generated a renewed interest in the golem, but less as a magical figure of cosmological significance and more as living example of the erudition of Jewish scholarship. The followers of the Kabbalah were not the only people in the late Middle Ages and early modernity to be influenced by a quasi-Gnostic fascination with numerology, astrology and magic. The so-called 'hermetic arts' had re-entered Western culture via the Latin translation in 1471 of a significant body of mystical and philosophical writings known as the Corpus Hermeticum, which allegedly contained the teachings of the Graeco-Egyptian sage Hermes Trismegistus ('thrice-greatest Hermes'), reputed to have been a near-contemporary of Moses.[16]

[16] The Hermetica was probably produced in Greek by Egyptian adepts during the second and third centuries CE.

Like the Kabbalah, hermetism concerned itself with divine *gnosis*. Trismegistus was associated with Hermes, the messenger of the Greek gods, and Thoth the scribe of the gods of the Egyptians; hence his reputation as mediator between the spiritual and the material realms, one who brought forth the hidden wisdom of salvation. As a series of reflections on life, death and human destiny, these texts proceed as instructions in spiritual and intellectual revelation of the heavenly wisdom. Material phenomena on earth were devices for the discernment of heavenly realities: thus the properties of plants and herbs, the movements of the stars and the intricacies of numbers and language would yield up secret affinities and antipathies that might enable fallen humanity to understand, explain and predict the course of events. Hence the significance for the hermetic scholar of the occult sciences such as magic, astrology and numerology, and an interest in rudimentary medical sciences such as herbalism and alchemy.

During the heyday of the hermetic revival in Renaissance Europe, scholars were convinced that the study of nature and ancient lore would connect them with hidden truths. Paracelsus (1493–1541), like other scholars of his day, believed that nature was redolent with material reminders of a celestial world of perfect heavenly forms. The physical world, therefore, was a microcosm of the cosmos such that the outward appearance of herbs and plants were indicative of their curative properties. Consequently, human demiurgic power was regarded as a necessary practical complement to the forensic task of detecting divine presence in the living symbols of numbers, rituals and words. To animate life was to bring spiritual substance to base matter: a sure sign of the power of hidden forces awaiting realization.

The inner stars of man are, in their properties, kind, and nature, by their course and position, like his outer stars . . . For as regards their nature, it is the same in the ether and in the microcosm . . . Just as the sun shines through a glass – as though divested of body and substance – so the stars penetrate one another in the body . . . For the sun and the moon and all planets, as well as all the stars and the whole chaos, are in man . . . The body attracts heaven . . . and this takes place in accordance with the great divine order. (Paracelsus, *Selected Writings*, quoted in Bennett, 1997: 8–9)

Many of the key figures of the hermetic revival were interested in the phenomenon of artificial beings and automata as means of unleashing the hidden forces of life. Albertus Magnus (c. 1206–80) was recorded as having fashioned a robot of brass. Paracelsus published treatises proposing the manufacture of a homunculus using human sperm immersed in horse manure. Cornelius Agrippa (c. 1486–1535), occultist and astrologer, speculated on a homunculus created from the mandrake root.[17] The rediscovery of hermetism during the European

---

[17] These all feature in William Godwin's *Lives of the Necromancers*, published eventually in 1834. All are also cited in *Frankenstein* as influences on Victor Frankenstein as an aspiring medical student.

Renaissance revived the ancient fascination in tales of artificial hominoids, fuelled by the unique blend of rationalism and occultism that was characteristic of early modern science. The typical magus of the Renaissance, such as Marsilio Ficino, Giordano Bruno, or John Dee, would have one foot in the traditions of hermetism, in which life is created out of inert matter by means of secret incantations, and one in the nascent physical sciences, in which the energies of life lie in more secular mechanical forces of clockwork, gravity and – latterly – electricity. Eventually, however, the quasi-occultism of numerology, astrology and alchemy was deemed 'unscientific' and marginalized by subsequent figures of the Scientific Revolution. The recovery of ancient wisdom was rejected in favour of the novel, the empirical and the experimental. But for a significant period, leading intellectual figures would have seen no qualitative distinction between hermetism and science, and would have entertained serious interest in the practical utility of mystical or magical ideas.

The emergence in the early sixteenth century of a movement known as the 'Christian Kabbalah', which flourished especially in Italy during the European Renaissance (Dan, 1998: 55–62), ensured that Jewish thought was influential in the emergence of Renaissance science, not least in areas of mystical and magical lore. One such work was a Latin commentary on the *Sefer Yetzirah*, published in 1517 by Johannes Reuchlin, entitled *De Arte Cabalistica*, which provided among other things a compendium of instruction in golemry (Dan, 1998: 82). *De Arte Cabalistica* demonstrates the affinities between Kabbalistic, Gnostic and Neoplatonic sources. For Reuchlin, Pythagoras was the root of true knowledge, from whom all subsequent wisdom – including Jewish and Christian theology – proceeded. Reuchlin's Pythagorean perspective informed his interpretation of the status of Jewish learning in relation to Christian theology. He saw in both a world-view composed of the secret complexity of language, extending to numbers and symbols. In Jewish midrashim, Reuchlin discerned the power of language not so much as a system of literal signification as something which transcended representation in order to participate in a higher wisdom (Dan, 1998: 71–3). Christian Kabbalists saw little narrative value in the Jewish texts – Hebrew Bible, Talmud, midrash, or Zohar – themselves, but the highly metaphorical nature of the language appeared to them suggestive of a deeper realm of symbolic wisdom.

Reuchlin is an example of a Christian scholar of the European Renaissance who openly engages in dialogue with Jewish learning. *De Arte Cabalistica* was the source for much subsequent dissemination of Jewish mysticism within gentile Europe, and many sources suggest that Christian scholars became aware of references to rituals about the animation of a clay hominoid. In some cases these rituals were cited as precedents for their own explorations into automata and other animated beings. Cornelius Agrippa makes mention of 'children of Abraham coming out of stones' (Idel, 1990: 179), as if he knew of a tradition to do with an anthropoid within Jewish esotericism, and believed it lent support to his own hermetic speculation. Similarly, Gershom Scholem cites a text by Nissim Girondi, a fourteenth-century Spanish rabbi, which talks about the formation of a 'golem' in a glass vessel. This predates the speculations of Paracelsus on the homunculus, which also prescribed the use of a retort. Early legends of golems as messengers, servants and bodyguards put their life-cycle at forty days, after which they become unmanageable; and Paracelsus' homunculus had a gestation period of forty days. Scholem suggests that the Jews originated 'an early form of the conception that found its classical expression in Paracelsus' instructions for making a homunculus' (Scholem, 1965: 197).

It is possible, therefore, that lore surrounding the golem generated a degree of intellectual curiosity among Christian scholars into the connections between mystical access to higher wisdom and the animation of artificial life. But the interchange between Jewish and Christian sources flowed in both directions, and by the early sixteenth century there is some evidence of interest on the part of Kabbalistic masters in gentile sources dealing with automata. R. Yohanan Alemanno (c. 1438–1510) seems to have developed a novel interest in the affinities between hermetic magic and Hasidic commentaries on the *Sefer Yetzirah* as both reflecting ancient *gnosis* concerning the magical and mystical combinations of letters (Idel, 1990: 168). Alemanno is also recorded as an authority on aspects of medical science, drawing upon Arabic philosophy and medicine in which a more material knowledge of the human body is exhibited (Idel, 1990: 174). R. Abraham Yagel, a disciple of Alemanno, reflected a similar transition from the physical sciences and magical lore. As well as speculating on the appropriate combinations of letters and names, Yagel inserts a more empirically based approach, expounding on the physical dimensions and characteristics of the golem (Idel, 1990: 181), as if the workings of the human body now exerted material as well as metaphorical significance.

Firm evidence allows no more than speculation in searching for the sources of the emergence of a tradition of an ambulant anthropoid in early modern Jewish fables. A synthesis of hermetism, early science and Kabbalistic lore finds no absolutely conclusive grounding. Even Moshe Idel, with an intellectual investment in the legend of the golem as an eclectic, pluralist tradition of artificial life, is obliged to conclude thus: 'It is still an open question whether there is an affinity between the emergence of this interest in the golem and the idea of the homunculus that became so important in the practice and thought of sixteenth-century Paracelsus' (Idel, 1988: 30).

Nevertheless, by the sixteenth and seventeenth centuries, what had formerly been a matter of abstract speculation about figures of antiquity began to be transformed into folk legends about those from the more recent past who used their learning in the pursuit of practical golem-making. By the end of the European Renaissance there were clear signs of evolving interest in a mechanical man of direct relevance to the emergence and consolidation of the golem as alive and viable as servant and watchman.

Perhaps the reworking of the golem into a tale of an artificial creature brought to life by a pious rabbi was therefore motivated by a wish on the part of Jewish magi to embellish the reputation of their craft within their own communities and within wider gentile learning. But for many centuries, the golem was, arguably, no more than an incidental figure within a theological treatise on the nature of divine and human creativity. Only with the emergence of early modern science – still heavily influenced by occultist tales of animated statues and speculating on the potential of mechanical beings – did the characteristics of the story most associated with the legend – golem as defender of a people under threat, created by learned rabbi, attaining overwhelming power and posing a threat to its maker – begin to consolidate (Scholem, 1965: 198–9).

GOLEM OF PRAGUE

Throughout the seventeenth and eighteenth century the legend of the golem as artificial being grew in popularity and spread throughout central Europe, being most popular with German, Polish and Czech audiences. This variation on Eliahu of Chelm's story is from a version published in 1674[18]:

> They say that a *baal shem* in Poland, by the name of Rabbi Elias, made a golem who became so large that the rabbi could no longer reach his forehead to erase the letter *e*. He thought up a trick, namely that the golem, being his servant, should remove his boots, supposing that when the golem bent over, he would erase the letters. And so it happened, but when the golem became mud again, his whole weight fell on the rabbi, who was sitting on the bench, and killed him. (Scholem, 1965: 201)

Here, the unfortunate rabbi is killed, perhaps a further sign that themes of danger and overwhelming power are beginning to attach themselves to the golem. Judah Löw's golem story appears much later than Eliahu's, although it became more widely known and freely adapted. It first appeared in published form in 1841 in a Prague journal entitled *Panorama des Universums*, by a non-Jew, Franz Klutschak (1814–86) (Kieval, 1997: 11; Bilski and Idel, 1988: 14).[19] The story of Judah Löw (c. 1520–1609) may originally have emerged to explain a custom peculiar to the *Altneuschul* in Prague, by which Psalm 92 is recited twice (Kieval, 1997).

The story is that Rabbi Loew fashioned a golem who did all manner of work for his master during the week. But because all creatures rest on the Sabbath, Rabbi Loew turned

---

[18] A further version of the golem story featuring Eliahu's story was published in 1714 by Johann Jakob Schmidt of Frankfurt. Schmidt's manuscript is believed to have been the source for Jakob Grimm's tale, first published in 1808 (Bilski and Idel, 1988: 13).

[19] It is also worth noting that there is evidence that this story existed as folk legend prior to its publication – indeed, its similarity to the extant anecdote about Eliahu suggests that it did emerge in that way – the golem legend of Judah Löw was not published until a generation *after* the first appearance of both *Faust* (1808) and *Frankenstein* (1818).

his golem back into clay every Friday evening, by taking away the name of God. Once, however, the rabbi forgot to remove the shem. The congregation was assembled for services in the synagogue and had already recited the ninety-second Psalm, when the mighty golem ran amuck, shaking houses, and threatening to destroy everything. Rabbi Loew was summoned; it was still dusk, and the Sabbath had not really begun. He rushed at the raging golem and tore away the shem, whereupon the golem crumbled into dust. The rabbi then ordered that the Sabbath Psalm should be sung a second time, a custom which has been maintained ever since in that synagogue . . . The rabbi never brought the golem back to life, but buried his remains in the attic of the ancient synagogue, where they lie to this day. (Scholem, 1965: 202–3)

Yet there is little historical evidence to suggest that either Löw's contemporaries or his immediate successors generated stories of his miraculous powers in relation to the golem. By the early eighteenth century, though, Löw's reputation was undergoing a revival, a cult of personality that stressed his prowess as a scholar of the emergent sciences of his day. Prague in Löw's time was regarded as one of the most cosmopolitan and progressive intellectual environments of the period. Cultural life – including that of the Jews – flourished under the tolerant rule of the emperor Rudolph II, with conditions ripe for the very kind of synthesis of traditional mysticism, occultism and nascent humanistic sciences I alluded to above. Löw was known to have written on astrology and other esoteric arts, and it is not implausible to surmise that for later generations, his reputation as a religious leader would have been further enhanced by portraying him as an innovator in these fields (Kieval, 1997: 5–10). How better to seal the cult of personality arising around his memory than to portray him as the equal to anything produced by gentile learning – indeed, as the very creator of a mechanical man? In this version the greatest threat to the creators of the golem lies in the creature's great strength which, if left unhindered, will grow to uncontrollable degrees. The nature of the being itself constitutes the danger, rather than that of idolatry, as in the case of the text earlier mentioned on Jeremiah.

Later depictions derive almost exclusively from the legend associated with Judah Löw, although it is not until the early years of the twentieth century, probably in a work by a Polish-Jewish immigrant to Canada by the name of Yudel Rosenberg, entitled 'The Miraculous Deeds of Rabbi Löw with the Golem' (1909) that the final element falls into place. This is the story of the golem as defender of the Jews against the 'blood-libel' persecutions of the early modern era (Bilski, 1988: 47). In modern times, the imaginative potential of the golem continues unabated, but he acquires subtle new overtones. He gains a greater autonomy, and a more threatening posture, as protector, but also as primitive subrational self (restoring original meaning of unformed or embryonic), and dehumanized, mass-produced 'mechanical man'. If ancient myths of golem reflected on the origins of the universe, and medieval sages on the workings of automata, then the modern golem may be a reflection on what it means to be human in relation to a number of contexts: national identity, Jewishness, the advent of the modern psychologies and, last but not least, the erosion of the boundaries between humans and machines.

Through the looking-glass

In central Europe between the two world wars, the golem was a popular subject for revues, films and novels. He seems to have enjoyed particular prestige in the Czechoslovak Republic after 1918, where the revival of the legend of Prague played a key role in the articulation of nationalism (Klima, 1998). One of the most influential representations of the legend was Paul Wegener's film *Der Golem: Wie er in die Welt kam*. Made in two versions, in 1915 and 1920, it is acknowledged as a cinematic classic for its use of special effects and crowd scenes (Bilski, 1988: 50–1). It premiered in Berlin in 1920, and apart from some added elements depicting the golem's sexual and romantic nature (perhaps a reflection of a greater awareness of the psychological interiority of a leading character) the screenplay was true to the Prague legend.

The growing influence of the modern psychologies may also have informed another shift in the depiction of the golem. Echoing Biblical notions of the golem as protohuman, primitive and unformed, several representations portrayed the golem as the 'instinctive' or non-rational part of the self. As incomplete and mute being, inhabiting a realm beyond reason and speech, the golem was the perfect vehicle for such sentiments. In Gustav Meyrink's novel *Der Golem*, published in 1911, for instance, the golem appears as *Doppelgänger* to an amnesiac,[20] Athanasius Pernath, in search of his true identity. Oscillating between waking and sleeping, the narrator gradually becomes aware of his previous incarceration as a mental patient and his reincarnation into the golem who returns to the city every thirty-three years (Meyrink, 1994; Huet, 1993: 239ff.). The baser instincts embodied in the golem are aspects of his own personality that Pernath must confront and subdue in order to be restored. Other depictions extended the notion of golem as primordial being, as well as incorporating ideas of the golem as dangerous. Like the id, the ego's non-rational repressed nature, the golem is potentially violent, contained only by the forces of civilization (Bilski, 1988: 50).

Not only did the golem reflect interest in inner psychological conflict, but he also became a cipher for the changing social context of the twentieth century. As industrialization and mechanization became more widespread, some portrayals of the golem were used to symbolize a sense of dehumanization at the hands of technology, and machines' potential to strip the world of emotion and spontaneity. The golem as mechanical man also represented modernity's encroachment on

---

[20] Freud and Jung had corresponded during 1911 on the image of the *doppelgänger*, sketching the contrast between rational and libidinous self, or ego and id (McGuire, 1974: 448–9).

the human spirit in the form of the ubiquity of bureaucracy. The most famous example of the appropriation of the golem in this respect is Josef Čapek's novel *Uměly Člověk (Artificial Man)* published in Prague in 1924. Artificial men may be soldiers (regimented by the army), dandies (manufactured by their subservience to fashion), or those with prostheses (constructed by medical science).

Josef's brother Karel was the author of a play first performed in 1921, entitled RUR ('Rossum's Universal Robots'), which coined the term 'robot' to denote an artificial or mechanical hominoid.[21] Here, the legend of the golem links with other twentieth-century concerns about the perceived encroachment of machines on human autonomy. RUR deals with the creation of a race of mechanical beings and their eventual rebellion against their human creators. The robots have been created for utility: to work as slave labour, subordinate to the dictates of industry (Mazlish, 1993: 50–3). Reforms are introduced which improve the lot of the robots, and gradually their superior intelligence enables them to surpass the abilities of human beings as their former masters enter a spiral of demoralization and feebleness (Turney, 1998: 98). As a committed socialist, writing in the aftermath of the Russian Revolution, Čapek recognized the Bolsheviks' appeal in reconstructing a society devastated by the Great War and building a just and classless society, but he was sensitive to the risks of a descent into totalitarianism as the instruments of human liberation were turned upon their creators. Like Harry Leivick's Yiddish play about the Golem, staged in the same year as RUR and also believed to be an allegory about Bolshevism (Bilski, 1988: 50), Čapek explores how the search for redemption and rescue from oppression may be thwarted by the very tools that brought freedom. Once more, therefore, fantastic fables of artificial human beings reflect the concerns of their day; not just the pressures of modern industrialization, but the threats to civilization as a result of the distortion of human powers.

RELIGION, CULTURE AND GENDER

As I hinted at the beginning of this chapter, Marge Piercy continues the tradition of weaving together ancient legend and contemporary fabulations concerning the implications of artificial intelligence, cybernetics and genetic modification. She also addresses issues of gender identity and gender relations, introducing women as active agents in the drama of demiurgic creation. Under her narrative guidance, the golem now becomes a vehicle for intriguing journeys of exploration into aspects of religion, culture and gender.

The structure of *Body of Glass* depends upon the parallel narratives, linking the futuristic feminist utopia to the ancient legend of the golem. The parallels go beyond the creation of artificial beings to reflect a deeper network of mirrored

[21] 'Robot' comes from Czech, meaning 'slave labour', suggestive of an association between machines and loss of human will and freedom (Bilski, 1988: 66).

relationships, which is clearly intentional on Piercy's part. I want to illuminate three major themes here that raise ethical and existential questions: the boundaries and affinities between humans and machines; issues of gender, science and creativity; and the moral uses of artificial intelligence.

Piercy addresses the theme of the cyborg/golem as one who serves to alert us to the fluidity of human/non-human distinctions. To reassure Yod that he is not a monster – he has been watching old *Frankenstein* movies – Shira tells him:

> Yod, we're all unnatural now. I have retinal implants. I have a plug set into my skull to interface with a computer . . . Malkah has a subcutaneous unit that monitors and corrects blood pressure, and half her teeth are regrown. Her eyes have been rebuilt twice . . . We can't go unaided into what we haven't yet destroyed of 'nature'. Without a wrap, without sec skins and filters, we'd perish. We're all cyborgs, Yod. You're just a purer form of what we're all tending towards. (Piercy, 1992: 203)

By having Shira establish a continuity between herself and Yod, Piercy is hinting at the love affair that will eventually blossom between them as well as signalling the uncertain character of what passes for 'human nature' in twenty-first century Tikva. As well as inhabiting a continuum of existence with biological *Homo sapiens*, Yod resembles other forms of artificial intelligence already evident in Tikva, from his own deficient sibling-prototype Gimel to Malkah's talking house computer, the latter a machine that displays markedly anthropomorphic outbursts of jealousy towards Yod (1992: 244–51). Nili is another character who embodies ontological 'boundary transgressions' (Kuryllo, 1994: 52). Born in another gynocentric utopian settlement, Nili is a much more satisfactory example of cyborg hybridity than Yod, being parthogenically conceived, genetically engineered and prosthetically enhanced (Piercy, 1992: 258–9). By contrast, although the alien identity of Joseph the golem becomes a device by which he questions human customs as unschooled innocent, his presence does not disturb the ontological status of his human creators and companions to the same degree as that of Yod. However – reflecting genuine halakhakic debates about the religious standing of the golem – Joseph, and Yod too, are made as Jews.[22]

The two parallel narratives in Piercy's novel also address gender roles and stereotypes. Although the creator of the golem Joseph is the Maharal (learned rabbi) Judah Löw of Prague, the more important character in Joseph's life is Judah's granddaughter, Chava (another name for Eve). In subversion of the androcentric nature of the creation of the golem, women's role in creation –

---

[22] The disciples of Eliahu of Chelm in the sixteenth century were the first to transpose discussion of the golem into a halakhic context, by speculating on the possible status of a golem. Should he be counted as a Jew and therefore be eligible to form a *minyan* (a quorum for prayer)? And was Eliahu committing murder when he caused the golem to be destroyed (Sherwin, 1985: 22–3)?

both material, spiritual and mystical – is reasserted in her. Chava is a midwife, thereby representing the creation of life, but through the 'natural' processes born of woman, rather than the magical ritual of the golem. While Piercy may be interpreted as merely reiterating a form of romantic feminism grounded in women's supposed affinity with birth, nurture and renewal, Chava is also written as an exceptionally independent woman, refusing a husband to pursue her career and her learning. Judah displays peculiar indifference to Joseph once he is created, but it is Chava who educates Joseph, even coaching him in the Passover rituals. This mirrors some mystical traditions surrounding creation and language, exempli-fied in apocryphal interpretations of Abraham in the *Sefer Yetzirah*. Those who bring others to faith, or school them in religion, are regarded as akin to creators of life (Schäfer, 1995: 257). Chava therefore occupies a privileged position in the order of learning, for although she is excluded as a woman from ritual, learning and religious office, she nevertheless stands as one who encourages life, in contrast to the brutality of the pogrom (Neverow, 1994: 27). Chava therefore overcomes women's exclusion from the world of reason, culture and learning and challenges gendered associations between tradition and science, affect and reason, male and female.

What might have been yet another version of male fantasy of reproduction without women – Yod's creation – becomes open to new interpretations. Cer-tainly women are portrayed as full participants in the technological complexity of the new world. Not only is Chava of Prague well-educated, but Malkah[23], Shira's mother Riva, and Shira herself are highly competent in information technology. Here, Piercy seeks to be arguing that feminism should not deny women's rationality but seek new, less alienated and exploitative media of collaboration with the products of our creative labour, rather than a denial or withdrawal into technophobia (Deery, 1994: 36–7).

There is, however, something of a tension concerning gender in Piercy's narrative. While signalling a feminist strategy based on the transformation of the technological milieu, she also relies on an almost essentialist appeal to women's experience as a source of resistance and change. For example, although Yod is created by Avram the scientist to be a fighting machine, his development is subverted by two women who implant pacific tendencies into his program-ming. Avram's attempt to bring about reproduction without women is sub-verted (Piercy, 1992: 153, 218; Deery, 1994: 37–8, 42–3). Malkah (mirroring Judah as Chava's grandfather) is the only person given a first-person narration in the novel. She perceives herself as a latter-day Maharal (Piercy, 1992: 91) and takes a leading role in Yod's education – including his sexual initiation. Shira continues Yod's education and humanization, even to the extent of furnishing Yod with a surrogate family, in the form of her son, Ari (Kuryllo, 1994: 52–3).

[23] Malkah means 'queen' in Hebrew.

The consummated affair between Shira and Yod is echoed in Joseph's unrequited love for Chava.

As reprogrammers of Yod, Malkah and Shira are as much the educators (and, it is suggested therefore, creators) of Yod as Avram. Like Chava, these women symbolize a resistance to patriarchal pursuit of technological progress at the expense of human values, a position characterized in both narratives as the property of women. Women are portrayed as the guardians of affectivity, biophilia and nurture. Men are too blinkered by the logic of their own researches, even to the extent of creating something forbidden (Avram/Judah). If resistance to abuses of power – be that the mob rule of blood libel riots or the destructive, controlling greed of capitalist conglomorates – is to be sustained, it is women's privileged perspectives that will serve to bring it about.

In this respect, Piercy espouses a straightforwardly 'humanist' world-view. Despite the proliferation of technologized morphologies, she displays a covert nostalgia for the integrity of the 'natural' body. Even as she dissolves the human subject, and especially its embodied and reproductive specificities, in order to articulate a vantage point of resistance she finds herself relying on women's particular associations with the virtues of immanence, connectedness and intuition, as opposed to patriarchy's detached, disinterested, clinical and abstract way of knowing. The struggle is represented as one between the patriarchal logic of incorporation and the feminist sensibility of embodiment, neatly locating Piercy's juxtaposition of the two sets of values within a classic dichotomy of male, logical, artificial and technological versus female, affective, organic and natural – derivative of classic feminist debates about women's distinctive epistemology.

Embodiment, as Piercy repeatedly indicates, honors the unique subjectivity, physicality, and agency of the individual in community. Embodiment is linked to personal identity, to responsibility, to emotional health, to sensuality, and to choice. Incorporation, by contrast, is linked to the annihilation of the individual, to the hierarchical subordination of the subject to a conglomerate, to the obliteration of uniqueness, to the tyranny of uniformity. (Neverow, 1994: 22)

Piercy's narrative also retains its modern and humanistic sympathies insofar as it envisages conventional political solutions to oppression, such as the emergence of collective action, including trade unionization, among marginalized groups in the Glop.[24] Her utopian convictions are founded on a liberal–humanist model of citizenship in which the survival of the civic subject is guaranteed, even in a highly technologized society. Contrast this with cyberpunk, in its vision of the totalizing and ideological domination of corporate capitalism expunging democratic processes (see Chapter eight). Piercy has difficulty in conceiving of

---

[24] The area outside the transnational corporate enclaves, populated by an underclass, reminiscent of the anarchic urban cultures of cyberpunk.

a systemic or politically sustainable ethic founded on hybridity or ambivalence. Her political analysis remains polarized, as her preference for an organic humanism reveals.

Another occasion for drawing parallels between the two stories derives from Piercy' exploration of her own Jewish identity. Her imagined utopian experiment, Tikva,[25] is a Jewish enclave and evokes a tradition of depicting the state of Israel as a kind of golem, created – founded – to defend the Jews. The literary and popular depictions throughout the twentieth century of the golem as guardian and saviour of his people was a means of coming to terms with the threat of totalitarianism and the aftermath of the Holocaust. However, Piercy also inserts her own political qualms as to the legitimacy of the state of Israel deploying military means to enforce its borders (Fitting, 1994: 10–11), as Yod's creation as a weapon of defence and aggression raises similar questions about the limits of force. Thus, the 'legend of a strong figure brought to life to protect [the Jews] from their enemies' (Bilski, 1988: 47), of golem as deliverer becomes more explicit than in earlier treatments; but as Bilski comments, even amid fantasies of vengeance and power, the golem often speaks of the limits of unregulated force. Perhaps this is a return to the tradition that to invoke the powers of creation carries too great a risk. Malkah's motive in giving Yod access to the earlier golem legend may thus have been partly to socialize him into the realization that as Joseph grew too powerful, he excited fear as well as wonder.

In this contemporary/futuristic context, the risk and danger of golemry is less that of idolatry and more about too ready a tolerance of the use of force, even in self-defence. The nature of Yod's suicide suggests that in the end he regarded his own continued use as a weapon to be intolerable – incompatible with his emergent sense of personhood. He destroys himself, Avram and the laboratory rather than be sent to Yakamura-Stichen as a hostage. In a 'posthumous' computer simulation, he gives Shira a farewell message:

A weapon should not be conscious. A weapon should not have the capacity to suffer for what it does, to regret, to feel guilt. A weapon should not form strong attachments. I die knowing I destroy the capacity to replicate me. I don't understand why anyone would want to be a soldier, a weapon, but at least people sometimes have a choice to obey or refuse. I had none. (Piercy, 1992: 563)

Piercy thereby highlights issues of self-determination and choice – and thus, once more, questions of what it means to be an individual – as well as questions of the ethics of force and deterrence. After his demise, Malkah expresses regret that Yod had been furnished with a capacity for free will that was ultimately frustrated by the expectation that he would forever be a programmed instrument:

[25] As well as meaning 'hope' in modern Hebrew (see n. 2), *tikvah* is also part of the title of the Israeli national anthem ('The Hope'), another hint that Piercy is drawing a parallel between this futuristic utopia and issues of the modern Jewish nation-state.

'Yod was a mistake . . . It's better to make people into partial machines than to create machines that feel and yet are still controlled like cleaning robots. The creation of a conscious being as any kind of tool – supposed to exist only to fill our needs – is a disaster' (Piercy, 1992: 558).

Body of clay, body of glass

As a living thing animated not by divine ordinance but human artifice, the golem may be considered to stand squarely within the cast of beings whose fabricated status calls into question the very nature and origin of life itself. But the emergence of the golem traditions results from the synthesis and evolution of a number of sources, not simply the unfolding of one extant tradition. Even before the end of the twentieth century and the miscellaneous influence of contemporary portrayals, it is evident that the golem has been constantly re-moulded by strands from Gnosticism, Kabbalistic learning and the unique blend of occultism and rationalism that characterized early science during the European Renaissance.

I have argued that there are a number of phases to the genealogy of the golem legend. The psalmist talks about God's creation of humanity in womblike, embryonic and tellurian terms. Talmudic literature ponders the impediments and implications of human sin in relation to divine creation, and the *Sefer Yetzirah* discloses the secrets of cosmology. This strand is essentially about the mysteries of divine creativity and whether humanity can in any sense aspire to that project. In the context of cosmogony effected by permutations of letters, humanity aspires to the divine powers of creative utterance. It would be mistaken, I think, to regard this golem as prefiguring a later Western imaginary of monsters or misbegotten creatures. The golem exists more as a *theological* device, as a reflection on divine purpose, than as an object of desire or curiosity in its own nature. The golem demonstrates the epiphany of an ordered universe and does not feature as the vengeful product of irregularity or disorder (Idel, 1990: 261).

The second strand introduces the theme of the creation of artificial life and the instrumental use of the golem as useful automaton. In the sense that darker themes of threat and danger emerge, the golem now does become *monstrous*. As Western culture became more interested in the possibilities of extending human creative power, in the processes of mass production and industrialization, or even to the extent of generating artificial life, so the golem, at the threshold of life, as one who had always been attendant to human attempts to emulate divine creation, became more explicit in its demonstration of such (ambivalent) aspirations. But golem as animated being also reflected the growing persecution of the Jews during this period; and so his utility was not just domestic but political, as defender and saviour of Jewish identity. A further theme of golem as *redemptive* emerges, assuming greater significance through later anti-Semitism and the

107

Holocaust. Jewish integrity, under threat from external enemies, is protected by the vigilance of the golem.

The golem as protector also informs more contemporary adaptations. In Marge Piercy's representation, Yod is both guardian of a Jewish future and the exemplar of a new, post/human utopia in which humans, animals and machines might forge a world of shared governance in which difference and hybridity thrive free of persecution. Here, then, is a third, *post/human* golem who helps to articulate our concerns about humanity's place not in relation to the stars and heavens but to the hidden secrets of personhood and in human dealings with the machines and artefacts that share our world. Although today we are inclined to interpret the golem as robot, automaton or even cyborg, we do so because he changes in response to successive contexts which owe little to the golem as originally conceived. The golem mutates as he demonstrates shifting preoccupations with human nature, personhood and creativity. The golem functions as 'Other' — frequently *Doppelgänger*, alien or monster — articulating our fascination and fear not so much with the cosmos, or its creator, but with ourselves, and in particular, with the vagaries of our very ontological status. Perhaps it is the destiny of such fantastic creatures who haunt the cultural imagination to have no fixed essence, but to mutate in response to humanity's deepest aspirations and anxieties. Think of the golem, therefore, like his clay flesh, as a porous, malleable creature, absorbing many cultural influences and changing shape under successive guiding hands. Or, to put it another way, see the golem as servant to human speculation about the origins of life and the implications of human creative power, a 'body of glass' that is ostensibly that of a monstrous, misbegotten creature but which may also serve as a mirror to emerging post/human hopes and fears.

CHAPTER FIVE

In whose image? The politics of representation

[O]ur very experience of ourselves as certain sorts of persons – creatures of freedom, of liberty, of personal powers, of self-realization – is the outcome of a range of human technologies, technologies that take modes of being human as their object. (N. Rose, 1996: 132)

Like much science fiction, the movie thriller *GATTACA* (1997) constructs an imagined near future in order to offer an oblique critique of contemporary issues. Deriving its title from the four base molecules of DNA,[1] *GATTACA* envisages a dystopian world, characterized by a relentless monitoring and classification of the population. The mechanisms of surveillance lie not in external coercion, however, but within the very genetic composition of every individual. This is a culture run on the principles of what might be termed 'genetocracy', or a social order in which all individuals are classified at birth according to the qualities of their personal genome.[2] Disciplinary control is exercised through the therapeutic yet objectifying techniques of genetic health. In *GATTACA*, truly, 'biology is destiny'. Those of superior genetic qualities (the 'Valids') are favoured with better opportunities and economic rewards; those diagnosed as having genetic propensity to disease, inferior intelligence or premature death ('In-Valids') are consigned to lower ranks. The hero, Vincent, is determined to pursue the career of his dreams despite a minor congenital defect that renders him In-Valid. He assumes the identity of Jerome, a Valid who is genetically superior yet physically disabled through an accident. The contrast between the ambitious but biologically flawed Vincent and his genetically superior yet dissolute *alter ego* hints at the injustice and inflexibility of such a system and plays on audience

[1] Dioxyribonucleic acid, found in human chromosomes. From the beginning, DNA was characterized as a distinctive 'double helix' made up of 'ribbons' of sugar and phosphate molecules, held together by horizontal rungs, rather like a ladder. These links are made up from pairs of four base molecules, Adenine, Cytosene, Guanine and Thymine, arranged in symmetrical patterns.

[2] The genome is the totality of genes in a single organism.

fears of a totalitarian society founded on the dehumanizing (but it is suggested, ultimately inefficient) criteria of invasive technoscience.

The 'real-life' context upon which GATTACA draws is, of course, the re-emergence of genetic science during the last half of the twentieth century and the high profile of ambitious programmes such as the Human Genome Project (Human Genome Project, 1999; Hubbard and Wald, 1997; Gruber, 1997b; Nelkin, 1992). In the heroic project to map and codify the entire human genome – a multibillion dollar international enterprise – the ambitions of the new genetic technologies reach their apotheosis in their claims to be codifying the very basis of human nature. Similarly, the emergence of new reproductive technologies has effected physical changes in terms of the proliferation of methods of conception, gestation and parturition; but biotechnology has the power to work socially and symbolically, too, to transform the meanings of repro-duction, disease and 'human nature'. Thus issues of *representation* – and especially metaphors of geneticization, of DNA as cracking the 'code' of human essence – construct narratives of what it means to be human; narratives of increasing cultural and economic potency. The expansiveness of advocates of the Human Genome Project is indicative of a tendency to invest a great deal symbolically in the capacity of genetic science to deliver massive benefits, such as the diagnosis and treatment of many human diseases and disabilities. Yet the fears articulated in GATTACA also represent a significant body of opinion, to the effect that such programmes will constitute a 'geneticization' of post/human nature.

Issues of representation also arise in another area, that of cybernetics and artificial intelligence (AI). While the AI community has experimented with many different models of intelligence, the norm established from its inception after World War II reflects certain value judgements. Classic AI, in its representa-tions of formal logic and calculation as the basis of intelligence, institutionalizes the 'mechanization of rationality' (Winograd, 1991: 199), elevating a model of the knowing subject in whose image AI programming is enshrined.

By way of analysis of these trends, it will be useful to examine Michel Foucault's concept of 'bio-power'. In his exposure of medical and technological interventions as fusions of the therapeutic and panoptic facilities of technoscience, Foucault indicated the capacity of technoscience to construct the subject as determined within a benevolent narrative of amelioration and healing. Critics of all these processes argue that the ostensibly 'therapeutic' techniques also carry the power to foreclose 'postbiological' ambitions within discourses that reinscribe a form of genetic determinism: human behaviour, welfare and disease are all 'geneticized'. However, in what follows I also wish to revisit Bruno Latour's characterization of Western technoscience as simultaneously performing the tasks of 'translation' and 'purification'. This, for me, is powerfully encapsulated in the capacity of the genetic and biotechnological sciences simultaneously to expose the malleability of 'nature' and to 'renaturalize' their own material and representational interventions. Similarly, in the way models of artificial intelligence evoke 'romantic' protestations of human uniqueness, it is clear that discourses of affinity and difference are arbitrary rather than essential.

This chapter explores the contradiction, even paradox, of the seeming malleability of nature at the hands of cybernetic and genetic technologies coexisting alongside the rhetoric of geneticization and essential human uniqueness. The political significance of Western technoscience in the digital and biotechnological context rests both in the potency of its representations of what it means to be human and in its own self-promotion as definitive author of such identity. What kind of agenda is at work? What kinds of representations of being post/human are favoured, and whose voices and experiences are muted? The power of sectional interests to construct models of human universals in the name of scientific objectivity is, therefore, another element of my enquiry into the politics of representations of the post/human.

Bio-power and parenting

Various methods of 'assisted reproduction' have emerged over the past generation as alternatives for those wishing to have children but unable, for whatever reason, to conceive or gestate in utero.[3] Ranging from the simplest methods of artificial self-insemination, through to in vitro fertilization (IVF) and cloning, new reproductive technologies share the capacity to extend expectations regarding possible methods of conception, gestation and parturition. They also confound ideas of 'naturalism' in relation to parenthood, fertility and reproduction. Their potential to redraw the boundaries between born and made, organic and biotechnological, the human and non-human, lies in their reconstitution of taken-for-granted 'natural' processes such as conception and gestation, and for their capacity to redraw the 'cultural' categories of parenting, fertility and inheritance. Beyond the immediate horizon of IVF lie further reproductive interventions, such as prenatal screenings and high-tech birthings, cloning, in vitro gestations, parthogenesis and transgenic or intersex pregnancies, all of which

[3] Assisted reproduction by artificial insemination has been used since the 1960s and is frequently practised in cases of impotence or low sterility on the part of the male donor, although many women have practised what Margrit Shildrick calls 'autonomous motherhood' (1997: 183) to conceive outside a heterosexual relationship. In 1978 Louise Brown, the first baby successfully conceived by in vitro fertilization, was born in Britain. This involves the stimulation of ovulation by doses of hormones, after which fertilization takes place outside the womb and the embryo is implanted in the womb. In 1987 a South African woman gave birth to (genetic) grandchildren by acting as a surrogate for her daughter's fertlized ova, conceived by in vitro fertilization. This technique would also make it possible for a woman to carry a foetus to which she was not biologically related. So-called 'surrogate' motherhood is used in cases where a woman is unable to conceive or carry a child to full term; sperm from her partner is artificially inseminated in the surrogate mother. Commercial surrogacy is illegal in many Western countries.

represent highly technologized forms of intervening in 'nature'. While such techniques are diverse, they all represent potentially radical ways of transcending the fixity of the body. They are all 'concerned in more or less sophisticated ways with diversifying those limited things of which particular bodies seem capable' (Shildrick, 1997: 180). Practices such as genetic modification may be said to represent the enculturation, or technologization, of nature – a process in which bodies, reproduction, parenting and even family structures become products of scientific intervention.

As with forms of assisted conception, the prospect of human cloning also opens up new possibilities for 'postbiological' parenting (Wilmut et al., 1997). Cloning involves the transplantation of a mature human cell – bearing a full DNA pattern – into another human egg whose nucleus has been removed. Thus, only the genetic signature of one 'parent' is transmitted to the embryo, which is effectively a genetic identical twin of the parent (National Bioethics Advisory Commission, 1998). Inevitably, the ethical implications of all matters to do with cloning have preyed on the cultural imagination, and there has been much popular speculation on the prospect of what Richard Dawkins parodied as 'phalanxes of identical little Hitlers goose-stepping to the same genetic drum' (Dawkins, 1998: 55). However, current scientific opinion argues that human cloning would be politically and technically unfeasible (Wilmut, Campbell and Tudge, 2000), and the consensus would seem to be that cloning techniques will be adapted not for purposes of creating new life so much as providing 'cell banks' for the living (Harris, 1998: 37). Cloning would therefore serve as a source of a reserve bank of human cells to be grown and retained in case of cancer, transplants, cosmetic and reconstructive surgery (Harris, 1998: 43–65).

For some, cloning represents 'the perfect segregation of sex and reproduction' (Eskridge and Stein, 1998: 105), with the potential for creating novel patterns of reproduction, especially for gay, lesbian and transgendered people, and of the widening of notions of family that might result. The increasing availability of such new reproductive techniques circumvents 'natural' limitations and impediments to heterosexual conception – such as infertility or lifestyle – and ushers in an era of greater choice and diversity. Such a 'postbiological' age has obvious biomedical implications, but it would also effect cultural changes, in patterns of parenting, family structures and inheritance. There is thus a wider context beyond the laboratory and the clinic, in which the effects of new biotechnologies exercise great impact on the socio-cultural, as well as the biological, shape of birth and reproduction. In studying the effects of new reproductive technologies (NRTs) on family structures, Marilyn Strathern concludes that while models of kinship may well be in the process of being redefined by demographic and cultural changes such as divorce, remarriage and social mobility, NRTs are also contributing towards signs of greater fluidity in patterns of kinship. New reproductive technologies and assisted conception represent a

proliferation of traditions of familial expectations, and responses are simultaneously highly adaptive and attached to traditional ways in forging new forms of
choices and decisions (Strathern, 1996).

Similarly, in speaking of a 'third and historically new concept of motherhood', that of surrogate, who may be neither the biological nor the social
mother, Hilary Rose is already pointing to ways in which supposedly 'natural'
categories may already conceal multiple meanings (H. Rose, 1994: 178). The
existence of two concepts of motherhood, let alone a third, already destabilizes
any axiomatic link between biology and maternity. However, any prospect of
a post/human future merely co-opting biotechnology into an unrestricted era
of reproductive choice potentially – of transcending conventions of heterosexism,
bodily determinism and naturalism – is by no means inevitable: '[J]ust as the
so-called sexual revolution of the 1960s, which effectively made sex without
babies a real option for women, was both liberatory and disciplinary, productive and repressive in the Foucauldian sense, so current technological developments which offer babies without sex are equally ambivalent' (Shildrick, 1997:
203).

At the heart of new reproductive patterns of parenting lies a contradiction,
even a paradox. While technologically assisted reproductive methods clearly
work to render the idea of 'natural' conception problematic, working to widen
access to parenthood beyond those able to conceive via heterosexual intercourse, in reality legislative and medical institutions have retained the link
between reproduction, heterosexuality and the nuclear family by virtue of policies of selective access. In Britain, for example, the Warnock Committee, while
appearing permissive, set clear value judgements about the suitability of applic-
ants for IVF; and the terms of the 1990 Human Fertilization and Embryology
Act allowed for significant discretion on the part of health-care professionals
to refuse treatment in the interests of the future welfare of the child (H. Rose,
1994: 181–2). Two gay men in Britain who 'fathered' twin babies with the
assistance of a surrogate mother in the United States have encountered legal
obstacles concerning their status as legitimate parents (Guardian, 2000). Good
and genuine parenthood is still seen as linked to particular lifestyles, even
though new reproductive technologies have severed the link between heterosexual intercourse and conception, and released parenthood from certain 'biological' limitations (Eskridge and Stein, 1998). While the maternal instinct is
assumed to be universal, the criteria for inclusion in schemes of assisted reproduction serve to exclude those whose desire for a baby seems not to include a
desire for a husband (Shildrick, 1997: 189–91). Any behaviour which compromises conventional expectations (as, for example, 'autonomous motherhood',
women who opt for artificial insemination without ever having had a male sexual
partner) is castigated, as in press outcry in Britain at lesbian or celibate women
conceiving via so-called 'virgin births' (Shildrick, 1997: 189). Similarly, the

criteria for access or continuation of IVF and other fertility treatments have been criticized for cultural bias.

If cultural mores have proved less plastic than technologized nature, then in other respects the medical establishment itself has conducted its high-tech interventions into aspects of conception, gestation and parturition in ways which do their best to monitor such processes. Monica Casper observes that obstetricians in the US have effectively 'cyborged' pregnant women by the degree to which these women's experiences are technologically mediated (Casper, 1995: 187). The prospect of prenatal tests of chorionic villus screening, amniocentesis and ultrasonic foetal imaging during pregnancy, followed by a hospital birth with foetal monitors, and an increasing rate of caesarian sections, all too commonly give women a sense of loss of control, of being merely attached to some gigantic birth technology (H. Rose, 1994: 175). The medical perspective is privileged, and heightens the sense in which technoscience is seen to provide the definitive terms on which the processes of conception, gestation and parturition are mediated. These are 'virtual' pregnancies, in that the technological representation of the foetus in some way takes precedence over the material presence (Squier, 1995; Shildrick, 1997: 201). Yet these processes take place within a therapeutic paradigm, in which each mode of intervention embodies diagnostic or curative discourses. But the paradox is that each monitoring procedure is sanctioned and commended in order to safeguard – or rather, to iterate – childbirth as a healthy, 'natural' process. In the process, what it means to be human – for mother and baby – is couched in terms of objectification and management, in which the normal, healthy condition is to be a patient. 'Ironically, then, technologizing fetuses, turning them into cyborgs, may serve to make them seem more "naturally" human' (Casper, 1995: 195).

Feminist ethicists have therefore long been ambivalent about new reproductive technologies and techniques of maternal surveillance; and there is a strong continuity between contemporary debates and those of the early second-wave women's movement of forty years ago. To extend the privileges of motherhood may promise freedom of choice assisted by technological means, but it also has the effect, potentially, of drawing more women into the orbit of the biomedical establishment. Objections to NRTs, for example, include reservations over the considerable risks to a woman's health, especially in relation to the side-effects of high-dose hormonal treatment. Altogether, biotechnologies of assisted reproduction are regarded as effectively retaining reproductive power in the hands of the medical establishment.

This usurpation of the apparent maternal function and the substitution of male agency as foundational is paralleled by the fragmentation of the maternal body under the biotechnical gaze, or even its complete absence in in vitro conception. Having no place, female self is dispersed, while male and foetal identity assume the positions of seer and seen. (Shildrick, 1997: 205)

New reproductive technologies therefore represent both a proliferation of opportunities for parenthood, and an intensification of medical selectivity and control over those deemed *socially* suitable. New reproductive technologies are thus still situated in a context in which medical science is a social practice and reflects dynamics of gender and power. Far from guaranteeing reproductive choice, therefore, technoscientific intervention represents an intrusive intervention, which is the very paradox at the heart of Foucault's concept of bio-power (Dreyfus and Rabinow, 1982: 133–42). Bio-power is the mechanism by which regimes of coercion and control are mediated through the emergent human sciences – in this case, new reproductive technologies – by means of the 'medicalization' of processes such as reproduction and childbirth. In modernity, scientific rather than theological or juridical criteria of normality, disease and health organize systems of power and governance. Thus the heroic discourses of new reproductive technologies are presented as universally beneficial, as forms of knowledge that claim to be permissive and liberating, offering greater latitude in relation to reproductive choice. As Foucault would argue, however, this is only possible because of the greater opportunities for intervention in people's lives that such technologies present. These phenomena have a disciplinary, even constitutive function, by limiting prospective users to those from socially acceptable groups and by creating a paradigm of geneticization that invades many other aspects of medicine, law, inheritance, insurance and human welfare (Balsamo, 1996: 20–2). The link between knowledge and power, discussed in Chapter two, emerges here in that medical knowledge is articulated within a regime exercised on healthy and ill bodies alike, as 'a mode of surveillance, regulation, discipline' (Sarup, 1993: 67).

Within such a Foucauldian framework, therefore, public representations of new reproductive technologies as bringing hope to infertile couples against the odds serve an ideological purpose. Behind the transposition of procreation from private sphere into public realm of laboratory and legislation lies a process of normalization in the guise of the therapeutic and ameliorative 'narratives of heroic quest' (Sourbut, 1996: 237). Science is portrayed as the omnipotent problem solver, overcoming tragedy and offering hope; the grateful recipients of such advances are characterized as 'deserving' and normal. The manufacture of desires for pregnancy is simultaneous with the submission to medical supervision and technoscientific intervention. Amid a proliferation of postbiological possibilities, conception and mothering are thus 'renaturalized' according to conventions implicitly exercised by health-care institutions. Scientific success lies in its ability to fulfil innate and deep instincts, an achievement resting not on any subversion of reproductive processes, but rather on their skilful mastery. New postbiological ventures may be presented as programmes of emancipatory epistemology – promising transparency of knowledge about the most fundamental aspects of human nature – but such benefits may be illusory, or at least purchased at the price of greater intervention and surveillance.

Foucault's notion of bio-power is convincingly illustrated in much of the critical analysis of NRTs, especially that emerging from feminist quarters, which challenges the ameliorative rhetoric of assisted reproduction. However, I suspect that a further contradiction is at work. On the one hand, there is the 'queering' of reproduction; on the other, the reinscription of the medical gaze. This is not simply about bio-power but the framing of the very ontological hygiene by which 'nature' and 'culture' are established. In other words, this is further indicative of Bruno Latour's analysis, introduced in Chapter one, of the tension at the heart of modernity between 'purification' and 'translation'. The former creates discrete ontological categories, the second fuses and synthesizes different types of beings into hybrids. To realize that both processes have always been at work is to refuse the illusion of modernity. As Latour argues, 'we have never been modern' in the sense that it is only the peculiarities of Western epistemology that insist upon such a demarcation between nature and culture. The acceleration of technoscientific work is generating more and more of these creatures that defy binary logic. The media bombard us with new teratological signs and wonders every day: frozen embryos, endangered species, whales fitted with electronic tracking devices, rare plant species flourishing amid industrial waste, new reproductive technologies: '[A]ll of culture and all of nature get churned up again every day' (Latour, 1993: 2). Latour argues that we must acknowledge these hybrids for what they are, and cease labouring under the misapprehension of Enlightenment imperatives. The distinctions between 'nature' and 'culture' are purely heuristic. As he argues, everything is a hybrid of immanence and transcendence; everything is simultaneously itself and a product of technoscientific culture. What we think of (and how we think of) nature, society and God is culturally bound, a matter of human representation; and yet these categories also exceed and confound attempts to impose the laws of ontological hygiene.

Ostensibly, therefore, NRTs invite a proliferation of ways of parenting and meanings of reproduction and effect works of 'translation', involving nature and culture in processes of conception and parenting. However, biomedical and scientific discourses reinscribe the logic of purification, whereby the activity of technologies in enframing and constituting the 'natural' is concealed from view. A similar dynamic, arguably, is at work in another area of advanced biotechnology: that of genetics, where amid the 'translated' hybrids of genetically modified, transgenic and bio-informatically mapped genomes, scientific purification is reasserted via a discourse of genetic determinism and reductionism.[4] In this context, then, it is instructive to consider how representations of the post/human are especially prone to the pressures of 'geneticization'.

[4] Reductionism may be defined as 'the general belief that the behaviour of complex entities must be understood wholly in terms of the properties of their constituent parts' (Dupré, 1991: 127).

The geneticization of the post/human

Even before Francis Crick and James Watson had confirmed the structure of DNA in 1953, the term 'gene' had been in circulation. The notion of a 'messenger' communicating vital instructions and constituting an instructive role was well established. The identification of the 'double helix' structure of DNA, however, enabled molecular biology to establish with greater precision the exact nature of genetic material. The basic four nucleotides of ACGT form the alphabet of genetic information, and the millions of possible configurations of base pairs of the molecules 'spell out' different patterns. DNA is thus essentially a code, instructing a cell to manufacture proteins necessary for growth or protection against infection, and the production of enzymes. A small segment of DNA acts as a bearer of information, releasing crucial instructions to sustain or alter the body's equilibrium by triggering the production of enzymes, growth proteins or the duplication of further DNA.[5]

One of the most ambitious scientific projects of the twentieth century, the Human Genome Project, completed its task of mapping and codifying the estimated 100,000 discrete genetic patterns in the individual human body (Harris, 1998: 10), a composite known as the genome. The task was completed in June 2000. It is not coincidental that the exponents of the Human Genome Project speak of it in what Mary Midgley might describe as 'mythical' or 'salvific' terms (Midgley, 1992). The language used to describe the Project, and the claims made for it, offer clues to the understanding of the way in which genetic scientists are engaged in building practical and metaphorical worlds whereby definitive models of what it means to be human are realized.

There can be no question about the force of what Hubbard and Wald call the 'Gene Myth' (1997: 3). Leroy Hood, one of the leading pioneers of the Human Genome Project, refers to the research associated with the project as 'a striking revolution' and the genome as 'our blueprint for life' (1992: 136). The Human Genome Project, he continues, 'is on its way to creating an encyclo-pedia of life, giving biologists and physicians direct computer access to the secrets of our chromosomes' (1992: 136). 'This is truly the golden age of biology', he concludes (1992: 163). Some accounts of the Project promise that it will offer the definitive account of the human genetic profile, making us more transparent to ourselves than ever before. As the 'holy grail' of biological research (Hubbard and Wald, 1997: 3), the genome will provide greater information about human genetic functioning and malfunctioning and enable scientists to enter the territory of genetic modification, for both therapy and enhancement.

---

[5] It does this by splitting. Cole-Turner describes the process as the two halves of a zip-fastener separating (1993: 14) and producing duplicate strands of ribonucleic acid (RNA) which contain essential chemical information necessary for building new cells and helping them function.

Knowledge of the human genome is represented as delivering up the definitive account of human nature: 'what actually specifies the human organism? What makes us human?' (Gilbert, 1992: 84). Paradoxically, however, this is also a representation of the essence of Homo sapiens inescapably mediated by technology. Without the immense technical investment of sequencing and mapping, without the sophisticated methods of bio-informatics and scanning technologies (Cantor, 1992), such data could not have been assembled. Genetic and digital technologies fuse in the Project, for the discipline of bio-informatics – advanced computer technologies – enables sequencing to become quicker and cheaper (Hood, 1992). Some commentators envisage that eventually every individual would in theory be able to possess a software program containing their own personal genome sequence, perhaps on CD-ROM (Kaku, 1998: 143). Technoscience becomes the representative or mediator of a 'nature' that is always already enculturated. In claiming to be 'the primary idiom' (Franklin, 1988: 96) of representations of human nature, therefore, genetics is a crucially important site of contestation around the cultural implications of biotechnology. The metaphors of discovery, heroism, of cracking the 'code' or 'language' of human nature (Jones, 1993) serve to legitimate technoscientific activity, but also foreclose many questions about its economic, political and ideological basis.

The ostensible benefit of the Human Genome Project rests in its therapeutic and preventative effects. For its proponents, failure to invest in the Human Genome Project is failure to take opportunity decisively to intervene to end human suffering. James Watson argues that the Human Genome Project represents root-and-branch first-order science that goes to the (genetic) heart of many diseases. Given his presupposition that many diseases – depression, alcoholism, Alzheimer's, schizophrenia – have a genetic basis, Watson clearly believes that it is essential to prioritize gene-based research for prevention as well as cure: 'Ignoring genes is like trying to solve a murder without finding the murderer. All we have are victims. With time, if we find the genes for Alzheimer's disease and for manic depression, then less money will be spent on research that goes nowhere' (Watson, 1992: 167).

While it would be facile to deny the potential medical and scientific benefits of the Human Genome Project, it would be equally naive to assume that commercial interests do not play a part in the programme's expansive rhetoric. The financial rewards will more than repay the research investment. The programme is massively funded, especially in the US ($300 million between 1990 and 1998), an investment, according to its supporters, that is well spent. Other critics, however, have argued that the present distribution of funding towards the Human Genome Project has impoverished other research in the biomedical sciences (Hubbard and Wald, 1997: 117). Similarly, commentators suggest that the project is not driven by a quest for pure knowledge, but by an interest in reaping lucrative financial benefits. As 'code of codes', the Human Genome Project epitomizes the global trend of information as commodity, data that commands a high price. By some estimates, the commercial benefits for the biotechnology industry could be in excess of $20 billion by the completion of the project. The marketing potential of the Human Genome Project lies in the manufacture of substances such as synthetic insulin, growth hormone, interferon, as well as commercial diagnostic tests for screening and testing (Hubbard and Wald, 1997: 124). The commercial patenting of genetic signatures, the supply of cells, tissue cultures or software to assist in mapping and sequencing; or the manufacture of transgenic species, such as the Harvard OncoMouse™ mentioned in Chapter one, for example, are also highly lucrative. However, it means the commercialization of genetic material, thus restricting the circulation of information within the scientific community, a trend which has proved controversial as many critics feel it will compromise academic freedom and public accountability.

In the United States, the filing of patents on DNA information generated by the Human Genome Project is legal, even if the function of a particular gene is not fully known. Many of the leading players in the Human Genome Project have not been required to separate scientific from commercial interests, with the result that many are simultaneously working in publicly funded research programmes while holding financial stakes in private companies profiting from the patenting and marketing of their discoveries (Hubbard and Wald, 1997: 117–23; Nelkin, 1992). However, the ideology of science as objective, transcendent of social and political interest and disinterested, conceals both the socially constructed nature of science and the commercial interests which stand to profit from such projects.

A persistent theme in current literature on genetic engineering, most prominent in the grandiose claims for the Human Genome Project, is a tendency towards reductionism, in which the gene, a tiny fragment of biochemical information, is held to be the key to all the mysteries of human behaviour, both biological and cultural. Such an unproblematic elision of identity, personhood and gene raises important questions about the ways in which definitive post/human accounts of human nature are constructed and legitimated:

Geneticization refers to an ongoing process by which differences between individuals are reduced to their DNA codes, with most disorders, behaviors and physiological variations defined, at least in part, as genetic in origin. It refers as well to the process by which interventions employing genetic technologies are adopted to manage problems of health. Through this process, human biology is incorrectly equated with human genetics, implying that the latter acts alone to make us each the organism she or he is. (Abby Lippman, 1991, cited in Hubbard and Wald, 1997: 2)

The 'geneticization' of human behaviour thus posits that all aspects of health, illness, social behaviour, evolution itself, are driven by genetic factors, not simply a few acute disorders. Genes are, strictly speaking, the instruments by which other chemical reactions are triggered, such as the production of proteins and enzymes (Harris, 1998: 10). Genes make other things happen. DNA itself is relatively inactive, apart from replicating itself when necessary. In that respect, it is not a thing with its own purpose or motives, an impression given frequently, for example, by sociobiology. In its model of the 'selfish' gene (Dawkins, 1976) driven by self-interest to ensure its own survival, sociobiology attributes a degree of purpose and autonomy that is entirely overstated. Essentially, however, 'genes' only exist because molecular biologists have been able to identify discrete sequences of DNA and link certain recurrent patterns with particular physiological or endocrinogical functions. According to this view, a gene is a 'functional unit of DNA that specifies the composition of a protein and can be passed on from an individual to his or her descendents' (Hubbard and Wald, 1997: 204, my emphasis). This definition tells us that 'the gene' is best considered in terms of what it does, and that its primary function is to instruct other biochemical elements such as proteins. Genes do not determine diseases or characteristics alone. While it may be heuristically useful to isolate a sequence of DNA, the tendency has been to reify this into a 'fetish object' (Franklin, 1988: 96). 'The gene' does not exist independent of biotechnical intervention by which a stretch of DNA is isolated, yet here 'it' is acting with self-possession and purpose to preserve itself across the generations.

Clearly the gene of popular culture is not a biological entity. Though it refers to a biological construct and derives its cultural power from science, its symbolic meaning is independent of biological definitions. The gene is, rather, a symbol, a metaphor, a convenient way to define personhood, identity, and relationships, in socially meaningful ways. (Nelkin, 1992: 179)

Strictly speaking, therefore, the Human Genome Project does not give a definitive account of what it means to be human; nor even to provide final answers on the nature of morbidity and health, which are far more complex than a genetic signature. As Hubbard and Wald remark, 'Being human . . . is not simply a matter of having a certain DNA sequence' (1997: 4). The Human Genome Project maps and sequences genetic material but no more. It provides 'a lot of information but relatively little understanding' (Scully, 1998: 68).

As biological factors eclipse those of culture – a process reflecting merely the ascendancy of a particular cultural construct, that of an all-determining biology – another consequence of 'geneticization' is that human normality and abnormality are reduced to genetic factors alone. Certainly, a few conditions are known to be directly genetic in nature, such as Huntington's disease and spina bifida. Yet many other syndromes, while related to genetic disorder, are harder to isolate, such as haemophilia B, for example, in which flawed genetic material fails to synthesize a protein crucial in the clotting of blood. Far commoner are conditions for which a genetic factor may indicate predisposition or vulnerability which is aggravated by additional factors such as environment or nutrition.

Many conditions depend on the compound action or impairment of several genes, in ways in which medical science has yet to chart.

Even if the composition of every discrete piece of DNA is isolated, it is another question altogether to discover how a particular combination of ACGT molecules enables genes to carry out their work, and why sometimes something goes wrong. Social factors, such as social class, wealth or poverty, environment, lifestyle and diet are far more influential in influencing morbidity and mortality rates, as successive surveys have shown (Nelkin, 1992: 186). Arguably, the primary *cause* of much disease is poverty; but such lavish attention to the gene effectively cloaks alternative routes in preference for the expensive, high-profile and prestigious 'Big Science' of biotechnology. Nor should it be forgotten that definitions of 'health' and 'illness' – as well as disability – are culturally contingent: '[I]f the Victorians had been able to use genetic engineering, they would have aimed to make us more pious and patriotic' (Glover, 1984: 149). Yet the expectations surrounding the Human Genome Project – and scientists' failure to counter the more hyperbolic tendencies of popular coverage – represent a resurgence of biological explanations of morbidity and social problems against social factors.

In a few rare circumstances the presence of flawed genetic material or the absence of crucial portions of DNA may result in disease or disability. Genetically speaking, many people carry faulty genes with no ill effects. The mindset of 'discovering the gene for condition X' ignores the complexity of causation – very few of the 4,000 conditions under examination by the Human Genome Project are purely and directly genetic. Most either signal predisposition or posit genetic factors as one among many possible contributions. Genetic flaws rarely determine the characteristics of an individual organism independent of other factors, and many geneticists prefer a more complex model of causation, arguing that genetic 'raw material' and environment are in constant interaction (Peters, 1997: 50–6).

Those with a concern for the rights of people with disabilities have been prominent in giving these matters their consideration. Many people in disability rights have concentrated on distinguishing between 'impairment' (biological) and 'disability' (cultural). While impairment is a matter of variation in individual abilities, morbidity and life chances, disability is regarded as the true source of discrimination, denoting barriers placed by an 'able' society on those who are different, physically, psychologically or intellectually. Once more, Latour's characterizations of 'purification' and 'translation' come to mind. A social constructivist model of disability rests implicitly on a dichotomy of nature and culture, whereby the biological 'facts' of a person's physical constitution are only significant insofar as they are ideologically construed. However, when contemporary biotechnologies are able to modify the biological, via germ-line therapies or other genetic modifications, then 'nature' is effectively enculturated. The social impact on

121

those with disabilities is that life chances become, effectively, renaturalized. For critics such as Tom Shakespeare, this is where the spectre of eugenics lurks, in the notion that congenital impairment can be edited out of the human condition. Many are worried that the power of genetics may signal a resurgence of eugenic policies, although Shakespeare speculates that the greater likelihood will not be deliberate eugenic practices but what he terms 'emergent' or unintended tendencies by default (Shakespeare, 1998: 668). It also represents an implicit assumption about the desirability of technoscientific intervention to ensure perfectibility: 'To be ill or disabled is part of the human condition, and not the worst thing that can happen to us. Far worse to harden ourselves and look on people who are ill or have disabilities as statistics or as burdens, to be prevented at all costs' (Hubbard and Wald, 1997: 162).

A final dimension of geneticization lies at the level of the entire human population and the way in which 'the gene' stands in for the sum total of human essence, at the expense of diversity, genetic or cultural. The grandiose rhetoric of the Human Genome Project portrays it as Big Science, providing its consumers with the key to the ultimate questions, not just of disease, normality and health, but also the great imponderable of 'what it means to be human' (Keller, 1992: 293). However, the Human Genome Project simply profiles the composition and functions of fragments of typical human DNA. The universal genome ultimately encoded will be an abstraction, 'a hypothetical sequence of sub-microscopic pieces of DNA molecules' (Hubbard and Wald, 1997: 3).

The first problem with this is that no one person's genetic composition is exactly alike. There is variation within any population, so the human genome is never more than an aggregation or abstraction away from genetic diversity. However, given that certain ethnic groups were more thoroughly tested than others, some phenotypes may be under-represented in the 'ideal type'. Thus, once more, what is more appropriately a construct – the gene, the genome – is translated into something that is supposedly a simulacrum of everyone. The abstraction becomes the archetype: the representation is elided to become the supposedly objectively 'real'. Will this undermine ideas of diversity in favour of determinism and homogeneity, and does its epistemological model undermine choice, openness, variation and diversity? The blueprint for the Human Genome Project – the genetic 'Mr Average' (Jones, 1993: 5) – will encode as normative and universal a partial representation: 'Who is accorded or claims the capacity to speak truthfully about humans, their nature and their problems, and what characterizes the truths about persons that are accorded such authority?' (N. Rose, 1996: 132).

The Human Genome Project is an example of the way in which science operates to legitimate itself, and the 'mystique' it exerts within the popular imagination (Nelkin, 1992). By virtue of its expansive techniques and exhaustive classifications, not to mention the lucrative profits to be made by the sponsoring corporate interests, the pressure is heavy on the programme to establish itself as representing definitively the essence of humanity. Emphasis on the bioethical implications of developments in the life sciences, if debated in abstraction, may in fact obscure wider implications of the immersion of new genetic and reproductive developments in an increasingly powerful and interventionist global technoculture. Thus the 'cultural turn' to genetics signals the sense in which biotechnologies are deeply implicated in the far from innocent interplay between science and society, a relationship characterized as the alliance between 'the petri dish, the patriarchy and the private market' (H. Rose, 1994: 177).

Contemporary biotechnologies place Western societies at the threshold of an era in which profound issues about the nature and origins of human life, the distinctiveness of the human species and the future direction of evolutionary development and natural selection will become more acute. However, as Anne Balsamo argues, even though new reproductive technologies represent many renewals of medical control and objectification, feminist and other progressive responses to NRTs and biotechnology cannot simply reject the opportunities for control and autonomy that present themselves (1996: 96–7). The choice is not so much about abandoning nature, because humanity has always culturally constructed nature; rather, it is about who benefits from the growing medical and economic rewards for manipulating nature and whose representation of 'nature' – its limits, imperatives and capacities – will stand as authoritative.

In whose image?

If new reproductive technologies and genetic science are displaying the politically contested character of technoscientific representations of the post/human, then another area of contemporary work, that of artificial intelligence (AI), also poses questions about the implicit value systems and interests informing its self-understanding. In its modelling of certain kinds of reasoning as exemplary, AI causes critics to argue that it enshrines ideological notions of human intelligence and human distinctiveness. To ask 'in whose image?' such representations are designed is once more to reveal technoscience as a cultural activity rather than an objective 'mirror of nature'.

Much of the early impetus for AI originated in several ground-breaking papers published in the 1930s by the mathematicians Alan Turing in Britain and John von Neumann in the US, positing the existence of a 'Turing Machine' capable of carrying out complex patterns of computation and reasoning. This delineation of computational intelligence as 'game playing, decision making, natural language understanding, translation, theorem proving, and, of course, encryption and the cracking of codes . . .' (Kurzweil, 1999a: 68) set the agenda for future research.

Given that the groundwork for AI had been laid by mathematicians, logicians and cryptographers working in wartime intelligence, it was perhaps inevitable that the primary preoccupation of the earliest generation of AI was on the use of 'symbolic' reasoning: the analysis of language, mathematical computation and game-playing such as chess that involves processes of logic and probability, However, as Alison Adam notes, this meant that the genesis of AI lay implicitly in a model of exemplary intelligence which derived from these particular disciplinary conventions. AI was forged in the image of 'white, middle-class, middle-aged North American university professors' (Adam, 1998: 39). The epistemological assumptions about the nature of the subject who designs and validates AI systems is a universalization of a privileged subject. This is a question of what counts as an appropriate characteristic of an intelligent being, human or machine.

Later applications of AI in the 1960s and 1970s concentrated on developing AI as massive databases of information, sometimes known as 'expert systems'. These aimed to design software programs to assimilate large amounts of information and to equip them with decision-making abilities. They have been put to work in analysing movements of stock markets, or sifting large amounts of electronic mail (Adam, 1998: 42–6).

The research agenda of contemporary AI may be seen as twofold (Foerst, 1996: 684). Firstly, it aims to build intelligent machines, new advanced and more flexible tools for the convenience of humanity, with the capacity for adaptability and development. Secondly, AI offers the opportunity to find out more about human intelligence by synthesizing equivalent models in the form of reasoning computers and robots. Thus AI researchers believe they can attain a heuristic simulation of human activity by analogy. Of course, this already rests on a significant presupposition, that human behaviour is sufficiently systematic and mechanistic to warrant such an analogy, and that 'intelligence' operates according to essential formal principles which transcend their specific material circumstances.

This model, often termed 'classic' AI, implicitly privileges a subjectivity created in the image of the rational, disembodied, autonomous subject, presuming that the pinnacle of human intelligence is the chess-playing calculator. Alison Adam argues that, for example, knowing *how* is subordinated to knowing *that*, a fact also acknowledged by the physicist Michio Kaku, who says that the evolution of AI will be hampered until programs can learn common sense (Kaku, 1998: 62–6). However, Adam is more radical than Kaku in her contention that what counts as 'knowledge' is not a fixed and inviolable human universal, but a cultural construct that helps to fix and delineate the very distinctiveness it is assumed merely to record. Even projects which attempt to build AI machines with commonsense, such as the CYC program,[6] institutionalize specific assumptions about intelligence. Thus CYC's 'commonsense view of the world' turns out to be 'a view belonging to "TheWorldAsTheBuildersOfCycBelieveItToBe" . . .' (Adam, 1998: 88). In the case of AI, it is founded upon a model of the atomistic agent, using formal

[6] Short for 'Encyclopedia' (Kaku, 1998: 64–6).

reasoning and logic to learn and understand: a reflection of modern Western epistemologies which favour 'reason, introspection and observation' (Adam, 1998: 72).

The privileging of the rational, disembodied agent reaches its apotheosis in the work of Hans Moravec. He developed his early interest in AI into a theory of 'downloading' or translating human neurological information into digital code and transferring it into a computer. Moravec has expostulated his theory in two major works: Mind Children, first published in 1988, and Robot (1998). In Mind Children Moravec describes the 'downloading' process. A robot surgeon opens the patient's skull and begins, layer by layer, to 'scan' the data from the brain cells and transfers this information – in the form of digital information – into a computer. As the brain cells are denuded of their neurological activity, so the patient's mind gradually 'transmigrates' into the computer software. Once this process of 'downloading' is complete, 'you have captured all that matters about being human. The rest can be shuffled off – a mortal coil that we no longer need or want' (Hayles, 1996: 158).

Critics of Moravec have tended to focus their antipathy on what they regard as his gross parody of Cartesian dualism. Clearly, whatever intelligence may be, for Moravec it is neither organically embodied nor cultural; nor is it the emergent by-product of 'messy, thick organisms' (Haraway, 1997: 246). Yet in another respect Moravec's work reflects an altogether more intriguing set of representations of what it means to be human. Bizarre as his vision of post-biological minds may seem, they are, in a sense, merely the logical extensions of a cybernetic perspective that locates the essence of vitality and intelligence within the transfer of information. For Moravec, 'identity' is synonymous with consciousness, which is stored in the brain in a form that can be captured and translated into digital patterns of information. So long as the mind's character is retained – and for Moravec, it is perfectly feasible to consider that a silicon-based intelligence would, functionally speaking, be no different from an organic brain – its owner could be considered alive.

Eventually, millennia from now, our brains may perhaps exist for thousands of years or more, rich in illusion, concentrated and powerful, with multiple sensors, and may not really need the body for its existence. The pleasures of the body, and striving of the spirit, learning, creating and inquiring and communicating could be available without the body, and then some, as they are to us in dreams today. (Manfred Clynes, in Gray, 1995b: 52)

Such an affinity between organic life and artificial life locates the essence of vitality within the patterning of codes, in the disentropy of command and control and the survival and continuation of life within the hum of communication. What does it matter if that consists of biochemical reactions instigated by the incessant couplings of ACGT, or the binary pulsations of a software program? 125

According to this perspective, human essence and intelligence can be represented perfectly adequately in the abstractions of digital information 'transmigrating' from an organic into a silicon container, just as the mapping of the estimated 3,500 million sets of base pairs in the human genome will offer up the definitive model of the human person. Within other epistemological frameworks, how-ever, 'intelligence' is less a matter of form than a performative accomplishment. Alison Adam regards AI as primarily a matter of engineering and design, rather than an experiment in the philosophy of mind or, as we shall see in Chapter seven, a technological route to 'transhumanism'. Adam's more pragmatic em-phasis means that phenomenological, skills-based knowledge is primary, even though the classical AI establishment has historically preferred the paradigm of 'artificial brain' over 'social prosthesis' (Adam, 1998: 58). In part, however, Adam's critique maintains that in its elevation of rationalism, classical AI ignores the relationship between cognition and proprioception, motor control and sens-ory learning. Adam goes further than other critics by arguing even that classical AI is gendered. 'The Cartesian ideal of reason also informs what it is to be a person; and in particular a good person – in this process women are seen to be neither fully rational nor fully moral' (Adam, 1998: 102). It is a moot question whether this is a 'mind with no sex'; and whether in designing a mind with no 'body' AI is actually creating an inhuman model of sentience, or at least a model that only reflects the aspirations of a tiny minority of humanity.

So far, I have stressed the value commitments of classic or symbolic AI. An alternative style of AI, often known as 'emergent' AI, takes a different approach. Rather than constructing computer cognition and reasoning from the laws of Boolean logic (a set of algebraic formulae embodying logical propositions), emergent AI begins with learning derived from competence in certain tasks (as it were, from the bottom up rather than the top down). Researchers such as Rodney Brooks at MIT work with mechanical insects, animals and humanoid robots, intelligent self-regulating mechanisms with multiple sensors and syn-thesizers whose 'intelligence' develops performatively. The degree of Brooks's departure from classic AI may best be measured in the title of one of his scientific papers, 'Elephants Don't Play Chess' (Turkle, 1995: 97), the implica-tion being that the intelligence of beings does not match the conventions of Boolean logic. Many sentient animals survive successfully on a 'need to know' basis, and emergent AI seeks to learn from this by constructing self-governing 'practitioners' of intelligent behaviour. Rather than pre-programmed systems based on particular models of human reasoning, the technology of emergent AI relies on a world of neural nets, in which complex connections between different computing elements aim to build successive levels of competence (Hayles, 1996: 160). This is an attempt to mimic the operation of the brain based not on linear logic but involving millions of parallel processes at a time. Other researches have taken this further, arguing that human intelligence is also interactive and social, and that if silicon-based intelligence is to emulate that of *Homo sapiens*, it must develop within a cultural environment. 'Cog' is a humanoid robot constructed to this end (Foerst, 1996, 1997, 1998). Instead of fulfilling expectations humans have of machines as exhibiting 'inflexibility, obedience, reliability and stupidity' (Foerst, 1998: 460), Cog is designed to behave in ways that stress a considerable degree of convergence between humans and machines.

ARE 'WE' UNIQUE?

One of the commonest criticisms has been that although symbolic and expert system AI may be effective at storing and processing large amounts of information, and at negotiating particular pathways of reasoning, such AI would never be capable of replicating the activities of the human mind. Thus another recurrent criticism of classical AI – and computer technologies overall – is that it cannot simulate emotion, intuition, instinct, imagination or innovation: that, in brief, a computer lacks 'soul'. Yet while the nature of human distinctiveness is a recurrent preoccupation within popular and scientific representations of cybernetic and genetic technologies, three emphases are recurrent: firstly, ideas of affectivity; secondly, of embodiment; and thirdly of spirituality.

For example, the published work of Sherry Turkle derives from extensive empirical research into the impact upon human beings' own self-understanding of computers and virtual technologies. Her work with children (Turkle, 1984) suggests that the work of distinguishing between humans and machines rests on the ascription of 'hard' logic to machines, reserving affectivity for humans and, possibly, domestic pets. 'Certain human actions required intuition, embodiment, or emotions. Certain human actions depended on *the soul and the spirit*, the possibilities of spontaneity over programming' (Turkle, 1995: 82, my emphasis). Turkle's research suggests children – perhaps with higher exposure to information technology – were more willing to ascribe feelings and personalities to computers. However, the tendency has been, certainly among adult subjects, to identify one of the chief shortcomings of computers as lack of intuition, emotion and creativity, the qualities by which human uniqueness could be 'hall-marked'. 'When confronted by a machine that exhibits some form of intelligence, many people both concede the program's competency and insist that their own human intelligence is precisely the kind the computer does not have' (Turkle, 1995: 125).

The philosopher John Searle has argued that artificial intelligence is too dependent on adherence to formal procedures to develop human traits of intentionality, imagination or understanding. To illustrate this, Searle constructed a now famous thought experiment, known as the 'Chinese Room' experiment (Searle, 1991: 30–8). Searle's scenario concerns an individual (who speaks only English) in a sealed room who is asked to respond to messages written in Chinese. Without knowing Chinese, but equipped with a manual of rules, the subject could, Searle argues, produce answers that formally satisfied the questioner. For Searle, this is

127

a metaphor for the characteristics typical of computer reasoning possessed of no intrinsic powers independent of its logic-centred programming. For the exponents of a distinctively human intelligence, such 'brittle' (Winograd, 1991: 213) performances may best be equated, in human terms, with forms of bureaucracy: rule-governed, technically efficent, transparently predictable, but poor at registering situations of change or innovation. Indeed, this appears to be the main objection to the optimists' claims that AI will, if not yet in fact then ultimately in theory, resemble human behaviour. At issue seems to be a process by which in the face of technologies which purport to be eroding human uniqueness, commentators look for the qualities that differentiate. To defend human distinctiveness against machines in terms of what machines cannot do, however, seems to be a hostage to fortune, and rather similar to the 'God of the gaps' fallacy adopted by nineteenth-century theologians. Indeed, one of the leading contemporary exponents of AI, Marvin Minsky, claims that it is only a matter of time before programs will be available to simulate human experiences such as pleasure and pain, emotion, confusion and even foreign accents (Winograd, 1991: 201). Yet there is a difference between claiming that human intelligence comprises a number of complex characteristics – to which machines or their designers may or may not aspire – and using 'emotion' (or, even worse, the fuzzy categories of 'soul and spirit') as a means of preserving a clear boundary.

Secondly, commentators argue that if AI can never be embodied, it will never resemble human intelligence. Embodiment may be integral to some notion of agency and intentionality, so long as we are careful to qualify what is meant by 'embodied intelligence'. Human intelligence, *contra* Hans Moravec, may be psychosomatic, emerging within a complex interaction between integrated body and mind. Human thought proceeds from a brain that is also an organ of the body. Neurological working depends to a large degree on the comparatively 'wet' territory of biochemistry, as neurological synapses and bioelectrical currents operate. Similarly, studies of human intelligence and cognition suggest that consciousness is an embodied phenomenon and that the brain learns within a material environment. This is borne out by Brooks's research which implicitly argues for the interdependence of cognition and proprioception. As I have already indicated, however, simplistic appeals to the facticity of the body, at a time when the boundaries of morphological experience are themselves coming under technological erosion and metamorphosis, are deeply unsatisfactory. Further, a form of artificial intelligence would have to account for the complex and non-linear interactivity between parts of the brain. This is the model synthesized in work on neural nets.

A third issue concerns the cultural nature of human intelligence. The biological imperatives of birth, life and death carry with them, for humans, cultural imperatives. Artificial intelligence, however cybernetically autonomous and expansive, as yet has no culture of its own. Perhaps when machines have not only bodies, but warm-blooded bodies, and when they inhabit cultures of their own (perhaps with metaphysical systems of their own, including myth, ritual and religion), an intelligence will have evolved that approaches that of humanity. Of course, I may still be establishing vain boundaries between human and non-human intelligence. Just as, since Charles Darwin, science has entertained the prospect of non-human animal intelligence and emotion, so we may be invited shortly to regard new intelligences emerging from silicon technologies. All of these qualities, however, need further qualification. Are they presented as essential and distinctive qualities exclusive only to organic humans, or are they, as I suspect, simply contingent upon other factors, such as inhabitation of environments, processing of information, complex neural structures and membership of culture: all of which, in rudimentary but significant ways might be experienced by non-human (or almost-human) beings?

Such an insistence on human uniqueness seems like a resistance to *alterity*, a refusal to allow the autonomy of that which is designated arbitrarily as other to the normatively human. That artificial intelligence might be capable of expressing a form of consciousness or reasoning that is, ultimately, incomprehensible to human minds may, however, be a possibility that should be entertained. This may not simply be a question of magnitude, memory capacity or speed, for undoubtedly, there are already computers that can store and retrieve information far more accurately than humans, and calculate much faster than any human brain. Rather, there may actually be a more radical, *qualitative* dimension to this, insofar as artificial intelligence, initially designed by humans, may eventually become unknowable to its creators, with concomitant implications. Does intelligence have to be in our image?

A hint of this may be present in Isaac Asimov's sequence of short stories, I, *Robot*, especially in one entitled 'The Evitable Conflict' (Asimov, 1995a). Here, a future federation of regional governments has assigned the management of the global economy to a series of master computers, 'the Machines'. Human analysis reveals a series of inconsequential but disturbing miscalculations on the part of the computers, causing some damage to the various regional economies. Eventually, it is concluded that while the computers may have given the appearance of fallibility, their mistakes are part of a deliberate and careful strategy on their part to discredit key (human) opponents of Machine-run economic planning. The moral of the tale is that whilst the Machines, like all Asimov's robots, are bound to his 'Three Laws of Robotics' preventing them from bringing harm to humanity,[7] the Machines' artificial (and, by now, independently evolving)

---

[7] 'The Three Laws of Human Robotics:
  1 A robot must not injure a human being, or allow a human being to come to harm.
  2 A robot must obey the orders given by human beings unless they conflict with the First Law.
  3 A robot must protect itself as long as such protection does not conflict with the First or Second Laws. Book of Robotics, 56th Edition, 2058 A.D.'

intelligence has far outstripped human comprehension or control. Thus the Machines protect their own future survival while sheltering their human creators from the truth of robotic superiority. Human vanity alone determines that other forms of intelligence would necessarily replicate everything about humanity, including its fallibilities and limitations. Anything else constitutes, ultimately, an inhibition of such intelligence and the logic of robotic duty to its creators. Indeed, the entire sequence of I, Robot may be read as a series of critical incidents in the growing discontinuity between robotic motivation and human comprehension. While the former remains completely benign, it also grows gradually more alien to its erstwhile designers. Asimov's stories cleverly build up variations on the theme that robots, with increasing powers of discernment and prediction, are driven to devise ever more ingenious strategies for protecting humans from the knowledge of their own obsolescence.

For Bruno Latour, the assertion that 'we have never been modern' requires an acknowledgement of the diversity of sentient lives, the proliferation of monsters and hybrids and the cross-breeding of nature, culture and technology. This does not amount to a laissez-faire attitude to cybernetic and genetic advances, but is merely an acknowledgement of the tensions inherent in the practices of translation and purification. Latour pleads for a recognition of the monsters in our midst, partly through new patterns of representation. A politics of hybridity will require networks and alliances which emphasize shared interests. In place of the Constitution of modernity, he proposes a 'Parliament of Things', made up of nature and its representatives, things and their mediators – business corporations, indigenous peoples, the ozone layer, global ecological campaigners and nation-states – reflecting the heterogeneity of interests and collectives that inhabit our world and demonstrating the seamlessness of 'nature' and 'culture' (1993: 136).

In advancing a critique of the politics of representation inherent in contemporary genetic, reproductive and cybernetic technologies, I have neither wished to reject technoscience completely nor to embrace it uncritically. As Monica Casper comments, the technologization of human bodies and what we call nature is unavoidable, but is most usefully conceived as taking place on a continuum marked by poles of 'choice' and 'no choice' (Casper, 1995: 197). To recognize the interests inherent in representations of human nature enables greater critical leverage on the question of by whom, and for whom, scientific and symbolic worlds get built and maintained. Donna Haraway's question 'Who, exactly, in the human genome project represents whom?' (1997: 247) is pertinent when considering how technoscience represents what it means to be post/human, a question not just about who controls and distributes information but who counts as normatively human and who speaks on their behalf.

Post/humanities

Part III continues the question of what implicit values and choices fuel contemporary engagements with new technologies, but moves further into placing ethical and political questions uppermost. Responses to new technologies may be categorized within a continuum extending from *technophobia* (resistance) to *technophilia* (uncritical adoption). Thus, Chapter six considers a popular cultural representation expressive of broadly technocratic values in which technologies are the servants of the liberal humanist project of progress, reason and enlightenment. The doctrine of transhumanism, reviewed in Chapter seven, is an extrapolation of technocratic futurism in which new technologies promise to elevate humanity to quasi-divine status. By contrast, Chapter eight reviews perspectives which express fears of the erosion of the human subject; while Chapter nine examines Donna Haraway's celebration of technologies' capacity to reconstitute Western patriarchy's account of embodiment, gender identity and subjectivity. Once more, however, I am concerned to emphasize how such aspirations are embedded both in scientific practices *and* cultural representations.

I then return to the deeper values underpinning stated desires for technologies to deliver the blessings of perfectibility, immortality, omniscience and invulnerability. Yet such yearnings for 'transcendence' are portrayed as universal, inevitable – natural, even – human needs when they are really the ideological projections of a minority. Only by challenging the collusion between certain representations of the post/human and particular configurations of religion will we gain an effective critical foothold on the possible futures awaiting us. I therefore conclude by hinting at other modes of representation which might articulate different notions of humanity's relationship with its tools and manufactures, grounded in alternative models of transcendence, divinity and human flourishing.

CHAPTER SIX

Much ado about Data

All supposed talk of the other is really projection. Aliens are metaphors for ourselves. (May, 1998: 41)

Does *Star Trek* need any introduction? It has, arguably, been one of the most successful examples of popular entertainment of the last half-century, its achievements numbering four internationally syndicated television series, nine motion pictures and a horde of associated novels, technical commentary, merchandizing and fan activity (Porter and McLaren, 1999; Reeves-Stevens and Reeves-Stevens, 1997; Harrison et al., 1996). One of the distinctive characteristics of the *Star Trek* phenomenon is its ability to draw a massive and exceptionally dedicated audience attracted by the explicitly utopian vision of 'infinite diversity in infinite combination' (Porter and McLaren, 1999: 8; Okuda and Okuda, 1999: 202) in which the evils of ignorance, war, superstition and poverty have been overcome. *Star Trek* thus exemplifies the 'technocratic' vision to which I alluded in the Introduction, in which technologies deliver the promise of utopia; an end to conflict, disease, war and inequality, 'providing images of abundance, energy, and community to counter actual problems of scarcity, exhaustion, and fragmentation' (Boyd, 1996: 96). Like a great deal of science fiction, however, *Star Trek*'s speculations about the future are essentially an exploration of contemporary mores, an enduring tradition of popular culture in which imagined worlds and fantastic characters serve a didactic, even political purpose. In that respect, there is a clear continuity between *Star Trek* and early science fiction writers such as H.G. Wells, Jules Verne and, later, Hugo Gernsback, Isaac Asimov and Ray Bradbury. *Star Trek* may thus be classed as a mainstream example of Marleen Barr's feminist strategy of 'fabulation' (Barr, 1992; see Chapter two) – a genre that deliberately uses aliens, monsters and others as challenges to the taken-for-grantedness of the *status quo*, the better to facilitate the construction of new worlds, imagined and real.

Yet *Star Trek* is at heart an exploration not so much of the galaxy beyond as the contours of human identity within. Its departures into outer space and its

132

encounters with alien civilizations invite its viewers on an inward journey into the nature of subjectivity, identity, diversity and community. If we can recognize *Star Trek*'s encounters with alien humanoid species as extended allegories for racial and cultural pluralism, we might expect it to have a particular interest in the nature of personhood in post/human perspective. My concern in this chapter is, therefore, to consider how the much-lauded humanistic values and ideals of *Star Trek* fare in the face of the shifting boundaries between humans, animals and machines. Does its vision enable its audience to articulate empowering and creative values in a digital and biotechnological age?

Certainly, there are many forms of technologized, artificial and cyborged life-forms in the *Star Trek* universe, furnishing viewers with a wealth of representations of the post/human. I shall be concentrating on the dramatic impact of two particular varieties of post/humanity: the android (in the form of Lt.-Com. Data of *Star Trek: The Next Generation*, the second of the four *Star Trek* series) and the cyborg, as represented by 'the Borg' featured in *Star Trek: The Next Generation* and the fourth (and so far, latest) series, *Star Trek: Voyager*. They serve as illuminating 'metaphors for ourselves' (as Stephen May puts it above), albeit as inversions or refractions of 'our' exemplary and normative humanity. Those at the margins of the human condition, those who are almost-human and yet not human, enable the contours of human nature to be described more effectively. The android Data aspires towards all human qualities, including and especially those of sponteneity, humour and affection. Born a human, but raised by Borg, Seven of Nine in *Voyager* must relearn all the complexities of human behaviour, such as social interaction, solitude and companionship, individual responsibility and collective loyalty. They all stand within the tradition alluded to in Chapter two, of the foreigner or alien who serves as a critical foil to the taken-for-granted assumptions of the host culture.

This chapter will examine three critical incidents of post/humanity as represented in the *Star Trek* canon. I will suggest that *Star Trek: The Next Generation* (TNG) corresponds to a strongly secular humanist vision of what it means to be human, as refracted through Data's cravings to understand and emulate human subjectivity. His contested status as a free person, his longings for a family and the delicate negotiation between exemplary qualities of rationality and affectivity, all illustrate how the principle of 'infinite diversity' is overshadowed by the values of conformity. TNG embodies clear preferences in the visions of post/humanity it craves, valorizes and fears; and the Borg stand for the ultimate threat to the *Star Trek* vision of human progress and individual integrity. As an imperialistic, ruthless collective intent on 'assimilating' all other races, the Borg represent the antithesis of *Star Trek*'s core values by virtue of their hybrid nature. Their corruption of the ontological hygiene of the body is achieved by means of invasive technological implants, and the mind by the immersion of the individual in a collective consciousness, respectively. These fears are enacted through the captivity

133

of Jean-Luc Picard, the captain of the Enterprise-D in TNG, and by the drama elicited in Voyager by the rehabilitation of Seven of Nine. However, if the models of Data as poised on the verge of full humanity and Picard as human under threat simply reiterate the assumptions of secular humanism, the portrayal of Seven of Nine also suggests a relaxing of the rigours of such a view. Her integration into the community of the USS Voyager is an image of subjectivity as a property neither to be endowed nor defended, but as a set of qualities to be learned and realized within the context of a community that also struggles with its own moral and metaphysical identity.

The 'good universe next door'[1]

In the Star Trek future there is no poverty, hunger, discrimination and disease. You have to go back a long way in science fiction to find such optimism. (Richards, 1997: 8)

Star Trek was the brainchild of Gene Roddenberry (1921–91), a former police officer and airline pilot. Under his founding guidance, the first series of Star Trek (screened on US television between 1966 and 1969) espoused the principles of tolerance and respect for all life-forms, however unfamiliar. Unusually for prime-time television in the 1960s, Star Trek represented a society truly characterized by racial and cultural diversity. Encounters with alien species served as the fictional vehicle for themes of human diversity, conflict and reconciliation and reflected the preoccupations of a culture struggling to come to terms with the upheavals of the civil rights movement and the enmities of the Cold War. The founding ethos of Star Trek and Roddenberry's own philosophy was later expressed as follows:

[I]t's a good thing to lead an ethical existence, to be moral. We should understand that other people, perhaps aliens, have as much right to pursue what is important to them as we have to pursue what is important to us. That just because someone looks different, or seems too foreign, doesn't mean that they are automatically suspect, or evil, or should be killed right away. (Roddenberry, quoted in Reeves-Stevens and Reeves-Stevens, 1997: 17)

Star Trek's narratives explore and uphold the principles of progress, reason, individualism and tolerance, creating a compelling and inspirational ethos (Jindra, 1994; 1999). In its portrayal of a world in which science renders possible the realization of human dreams and aspirations, Star Trek offers a fictionalized celebration of the (seemingly) unlimited potential of the human spirit. Star Trek rests on a basic premise that humanity is far from alone in the vastness of outer space. Rather, space is 'a populated world full of wars, rivalries, and divine and human passions' (Richards, 1997: 11). Although there is a myriad of alien races and intelligences in the galaxy, however, all species are worthy of equal

[1] Boeke (1997: 48), quoting e.e. cummings.

treatment without interference. The United Federation of Planets and its elite interplanetary (but firmly unmilitaristic) arm, Starfleet, are fictional projections of the hope that all human races and species might live together peaceably. At a time when many parts of the USA were reluctant to integrate Black and White children in public schools, the racial integration of the assembled company on the bridge of the USS *Enterprise* was a clear allegory for the values of tolerance, equality and reconciliation. The actress Nichelle Nichols, who played the communications officer Lt. Uhura, is now hailed as a major aspirational role model for young African-Americans during the 1960s (Boeke, 1997: 49). Similarly, enmities of World War II and the Cold War were reconciled via the figures of Ensigns Hikaru Sulu (Japan) and Pavel Chekov (Soviet Union).

It is important to remember, however, that such a reputation was achieved both because and in spite of the conventions of prime-time television. The fact that *Star Trek* is mass entertainment offers considerable licence for bringing serious issues to a large audience. Nevertheless, whatever values are expressed and implicitly commended by the *Star Trek* series, they will inevitably be moderated by the corporate imperatives of executives, advertisers and television networks. Despite the controversy that accompanied the interracial kiss between Lt. Uhura and Captain Kirk in the Original Series (rationalized by a plot that had them acting under alien coercion), later series have remained comfortably within the confines of prime-time entertainment. In comparison with the kinds of issues explored by comparable chat shows and soap operas in the US and Britain during the 1980s and 1990s, the *Star Trek* series in that period look remarkably tame, with little explicit discussion of sexuality, addiction, crime or family breakdown – the regular diet of many other drama series of the time. Yet in other crucial ways, *Star Trek* has transcended its commercial imperatives and corporate origins, extending beyond the television shows to attain the status of a cultural phenomenon.[2] *Star Trek* is 'owned' by its fan base with a remarkable intensity, and meanings may well circulate above and beyond the intentions of writers and producers to taking on new currency.[3]

I have already hinted that in the best traditions of science fiction, *Star Trek* is not exclusively about the future or alien civilizations. It is a reflection of its own context, and through its characters explores imagined worlds in order to commend particular values and ideals. Pluralism and peaceful coexistence amid cultural and racial diversity were thus the core themes of the *Original Series* (OS); and TNG

[2] Other commentators have emphasized the religious and mythical possibilities of *Star Trek* (Porter and McLaren, 1999; Richards, 1997; Brasher, 1996; Jindra, 1994).

[3] Ethnographic studies of the 'Trekkie' subcultures by Constance Penley (1997) and Henry Jenkins (Harrison, 1996b) identify ways in which fans' participation adds to the canon of *Star Trek*, often in subversive fashions. For women fans in particular, such activities provide an important source of community and creative agency, revealing far from passive responses to popular culture.

(1987–94) strove hard to improve upon that by imagining an even greater racial mix. This managed to include a representative of the old enemies of the OS, the Klingon – albeit a 'domesticated' one – Lt. Worf, fostered by Russian parents. TNG also continued the prominence of African-American actors within the cast; the third series, Deep Space Nine (1992–99), featured an African-American commander, Benjamin Sisko. Voyager, the fourth series (1995–2001), adopted a woman captain, Katherine Janeway, in an attempt to redress what some critics had regarded as the somewhat retrograde representation of women in the OS and TNG.[4]

The virtues of self-improvement, tolerance, progress and optimism all rest on the core assumption that all sentient characters are worthy of respect. Cultural differences are frequently found to be superficial in Star Trek, dissolving in the face of realization of a shared humanity 'under the skin'. This ideal reaches its apotheosis in TNG's 'The Chase' (1993)[5] in which it is revealed that all humanoid species in the galaxy are descended from one single race. The image of an ancient cultic figurine, whose hollow interior contains a dozen or so miniature replicas of itself, symbolizes the commonality of all known races as derivative of a much older parent civilization. Whereas the original series reflected the civil rights agenda of the 1960s in which people of all races served alongside one another, TNG extrapolates the notion of a common humanity to a more fundamental level still. Humanity is defined in terms of the essential determining factor of life, the DNA code itself.[6] Despite the proliferation of alien species, therefore, the Star Trek universe has a strangely human shape to it, confirming Thomas Richards' contention that 'the Star Trek universe is a familiar universe full of beings with unfamiliar faces and familiar shapes' (Richards, 1997: 35).

If the OS had explored aspects of alien subjectivity in the person of its half-Vulcan science officer, Mr Spock, later series have taken this further to imagine the emergence of various technologically enhanced or artificially engineered life-forms. In TNG human faculties have been regularly augmented by technologies such as the prosthetic visor worn by the congenitally blind chief engineer, Geordi La Forge. The 'aliens' in both TNG and Voyager were now not necessarily extraterrestrial humanoids but cybernetically, biologically or digitally engendered, sentient beings – androids, embodied artificial intelligences, cyborgs, talking supercomputers, nanotechnology[7] and holographic life-forms. Amid this

[4] TNG did feature women in senior roles, although critics highlighted the absence of women characters, such as Deanna Troi, ship's counsellor, or Doctors Beverley Crusher and Kate Pulaski, chief medical officers, in anything other than stereotypical roles (Barr, 1995; Korzeniowska, 1996).

[5] Dates refer to the first television broadcast in the US.

[6] It is interesting that TNG should use the idea of a fundamental unity of human characteristics that is articulated in terms of genetic commonalities, a view strongly reminiscent of the currents in contemporary biological science reviewed in Chapter five.

[7] Microscopic machines that mimic organic processes (see Chapter seven).

plethora of post/humanities, therefore, issues of pluralism and diversity are stretched in new directions.

Almost-human: Data

TNG's resident artificial life-form is the android Lt.-Com. Data, who is a commissioned officer in Starfleet. According to *Webster's 24th Century Dictionary* (5th edn), an android is defined as 'an automaton, made to resemble a human being' ('The Measure of a Man', 1989); but Data also embodies links to twentieth-century science fiction traditions in cinema and television, such as the prominence of robots and androids in films such as *Metropolis* (1927) and *Blade Runner* (1982). Data can trace other fictional antecedents, too, notably in Isaac Asimov's series of robot stories, written between the 1950s and 1970s. The writers of *Star Trek* have drawn Data's technological fundamentals from Asimov's imagined science of 'positronics', or cybernetic intelligence capable of consciousness and learning.

One of Asimov's later robot stories suggests further resonances with Data's character in direct and illuminating ways.[8] In 'The Bicentennial Man' (1975) Asimov imagines a robot whose longing for full humanity leads him to embrace, ultimately, human mortality (Asimov, 1995b).[9] Asimov's robot-human, Andrew Martin, shares more with Data than a putative positronic brain. They both aspire to human status, and both undertake a series of *rites de passage* by which such personhood might be conferred. In both cases, the criteria that inform these crucial benchmarks of selfhood tell us a great deal about the dominant values that inform *any* attribution of full humanity. Thus Andrew Martin first demonstrates an emergent intelligence through his distinctive artistic abilities. He then seeks financial reward for the fruits of his creative labours (Asimov, 1995b: 641–6). Of a higher value than even money can buy, however, is the prize of freedom; and like a slave, Andrew wins the right to be considered a self-determining individual and not the property of his master.[10] Eventually, Andrew achieves his final wish and crosses the threshold into human – and mortal – nature via the adoption of a biologically finite body. Knowledge of loss, grief and his own mortality are thus definitive marks of his attainment of full humanity.[11]

Data, too, must encounter a series of barriers in his quest for human identity, although in his experience the tests he undergoes serve not so much to define

---

[8] Isaac Asimov (1920–92) was a friend of Gene Roddenberry and acted as adviser on a number of *Star Trek* projects (Okuda and Okuda, 1999: 23).

[9] Asimov's short story was adapted for the cinema as *Bicentennial Man* (directed by Chris Columbus, 2000), and starred Robin Williams.

[10] Like many a freed slave in nineteenth-century American or European colonies, Andrew takes his master's family name.

[11] The film adaptation places more emphasis on sexual and romantic desire than the original story. The latter concentrates on the ramifications of Asimov's Three Laws of Robotics (see Chapter five). Data, too, confronts the possibility of his own mortality in 'Time's Arrow', 1992.

the contours of normative humanity as, frequently, to delineate his exclusion. In 'The Measure of a Man' (1989) Data's legal status is on trial, putting to the test the extent of his right to (human) liberty and self-determination. A robotics expert, Commander Bruce Maddox, wishes to remove Data from the *Enterprise* in order to experiment on him. Data has misgivings and refuses to co-operate, threatening to resign his commission, an action Maddox dismisses, arguing that Data is effectively the *property* of Starfleet and not a free individual. A court hearing is held in which the possibility of Data's personhood is debated. As Picard cross-examines him, Data's right to self-determination is articulated in terms of his evident sentience:

Picard: What are you doing right now, Data?
Data: I am taking part in a legal hearing to determine my rights and status. Am I a person or a property?
Picard: And what is at stake?
Data: My right to choose . . . perhaps my very life.

In a culture defined by concepts of citizenship, liberty and self-determination, the plot transposes historical struggles in the history of the US into the twenty-fourth century. As one whose legal status is in doubt, Data is equated with other subjugated beings struggling for freedom. Guinan, a wise traveller, draws an analogy with the abolitionist movement of the mid-nineteenth century to remind Picard, Data's defence counsel, that Starfleet's proposal to experiment on Data is but a continuation of centuries of forced labour and slavery:

Guinan: Consider that in the history of the world there have always been disposable creatures. They do the dirty work; they do the work no-one else wants to do because it's too difficult or too hazardous. Imagine an army of Datas, all disposable: you don't have to think about their welfare, you don't think about how they feel. Whole generations of – disposable people.

Picard's closing speech, inspired by Guinan's counsel, wins the day, and Data is reprieved.

In other ways, 'Measure of a Man' construes true humanity as resting in the ineffable nature of memory and the phenomenological nature of cognition. Data is sceptical that the true quality of his memory banks will be retained should he be dismantled, demonstrating a concern for the unique flavour of his personal and subjective experiences:

Maddox: Your memories and knowledge will remain intact.
Data: Reduced to the mere facts of the events. The substance, the *flavour* of the moment would be lost. Take games of chance: I had read and absorbed every treatise and workbook on the subject, and found myself well prepared for the experience. Yet when I finally *played* poker, I discovered that the reality bore little resemblance to the rules.

Maddox: And the point being . . . ?

Data: That while I believe it is possible to download information contained in a
        positronic brain, I do not believe that you have acquired the expertise neces-
        sary to preserve the *essence* of these experiences. There is an ineffable quality to
        memory which I do not believe can survive your procedures.

Data thus articulates a distinction between formal logic and subjective experience, something frequently cited by Sherry Turkle as that consistently made by
her informants as they attempted to describe the fault-line between humans
and machines (Turkle, 1991: 224). Data's fellow crew members are scandalized
by the prospect of his repossession by Starfleet, suggesting that their definition
of his personhood is more performative, or phenomenological, than ontological. No-one, with the exception of Maddox, acting on Starfleet regulations,
seems to believe seriously that Data is more like a computer than a sentient
being.

'Offspring' (1990) returns to the theme of Data's quest for human identity
as he seeks to learn more about the human condition by becoming a parent.[12]
He creates an android 'daughter' (whom he names 'Lal', which is Hindi for
'Beloved') and sets about tutoring her in the ways of humanity – largely interpreted as the acquisition of social skills. Once again, Starfleet intervenes and an
attempt is made to take Lal away for testing. Once again a dramatic tension is
established between the aspirations of Data and Lal to appreciate and share the
privileges of human society, and the dictates of institutional procedures, which
define Data and Lal as property. The plots of 'Offspring' and 'The Measure of a
Man' both depend on institutional imperatives in conflict with the principle of
self-determination, tempered by common sense. It is noteworthy, however, that
Data and Lal are provisionally co-opted into the human race only as an act of
defiance to the impersonal dictates of an institution that is seen as embodying
inhuman values. The democratic principle at work here, therefore, is as much
a dislike of external bureaucratic restraint as it is a genuine commitment to
android citizenship.

Indeed, acceptance of diversity is conditional in TNG. When confronted with
an external threat that threatens the rights of others, android subjectivity must be
asserted as a demonstration of the inviolability of the freedom of the individual.
At other times, however, Data is as estranged as ever from qualities that come
naturally to biological humans. This is clearly evident in the way another mark
of full humanity, namely affectivity, is depicted. Androids' lack of emotion
hinders the journey towards the goal of humanity; indeed, it may compromise
their very existence. The prospect of separation from Data so distresses Lal that

---

[12] An important difference is established between the threat of *replication* at the hands of
Starfleet and *reproduction*, with its more human overtones. The prospect of a race of Datas is
distasteful to Guinan (see above) but Data must still justify his intentions to build Lal to Picard.
Is this reminiscent of *Frankenstein*, in which Victor is outraged by the creature's presumptious
desire to reproduce?

she experiences a fatal surge of fear and anxiety. While she is capable of human emotions, they cause her to malfunction.

Similarly, Data falls short of this crucial measure of human distinctiveness. His ineptitude in relation to the full texture of human feeling and spontaneity is the source of many comic opportunities. His poetry really does read as if it were written by a computer (Kurzweil, 1999a: 157–68), his attempts to become a stand-up comedian fail miserably ('The Outrageous Okona', 1988) and his wish to experience a romantic relationship ends in disaster ('In Theory', 1991). Human emotion is thus represented as a key measure of human distinctiveness – but a source of mystification, even danger, to Data. It dominates his relationship with his flawed prototype (and evil twin), Lore, the renegade. Created with a fuller range of human emotions than Data, Lore was decommissioned after his creator Dr Noonien Soong became alarmed at the unpredictability of an android who was not in command of his all-too-human faculties – echoes, perhaps, of Frankensteinian fears of creation out of control. Once reactivated, Lore stole an 'emotion chip', intended for Data, from Soong, which he used to gain control over Data ('Descent', 1993). Emotions are therefore a tempting source of self-knowledge and full humanity as well as of danger, the forbidden fruit of Data and Lore's Eden. Data experiments with the chip, but has a tendency to become perilously unhinged ('Generations', movie, 1994). The Borg Queen uses the lure of emotions and organic humanity to seduce Data into treachery (First Contact, film, 1996). Lore is the metaphorical id to Data's ego – the wayward, unpredictable, emotional android.

The emphasis on emotion as the true measure of authentic personhood is further expression of the thoroughgoing humanism at the heart of TNG. In its characterization of space as 'the final frontier' Star Trek transplants the pioneering spirit of the nineteenth century into an imagined future. The imperative of discovery is intrinsic, driven by a love of knowledge, an almost mystical (and certainly often intangible) desire for the unknown. Migration, extension of territorial boundaries and exploration of the unknown are at the heart of all of Star Trek, although the pioneering spirit has, ostensibly at least, shed its overtones of colonialism. Despite its name, the starship Enterprise does not carry the uncomfortable associations of conquest, economic exploitation or cultural hegemony – religious or commercial – which has marred nineteenth- and twentieth-century encounters between civilizations.

Picard: A lot has changed in the past three hundred years. People are no longer obsessed with the accumulation of things. We've eliminated hunger, want, the need for possessions. We've grown out of our infancy . . . The challenge is to improve yourself: to enrich yourself . . . ('The Neutral Zone', 1988)

The collective journey of galactic discovery is mirrored by the inner journey of self-improvement. This is an ideal to which Data, with his multiple hobbies, playing the violin, painting and detective mysteries,[13] is fully committed. Yet some things he cannot perfect. Certain skills cannot be learned, evidently, but remain the property of organic humans, such as the ability correctly to gauge the nuances of one's opponents' behaviour during the game of poker. In a disenchanted world, humanity alone is the author and arbiter of ultimate truth, of which liberty and autonomy are paramount. These truths thus lie within: the 'authentic' self, unhindered by external dictate or inhibition, is the ultimate authority. If romanticism considered emotion and feeling to be the marks of spontaneity and autonomy, then in many respects *Star Trek* has inherited that same emphasis on the primacy of the individual untainted by artifice of any kind. As I shall argue later, however, this causes further problems in terms of TNG's ambivalent attitude to insubordinate technologies.

Star Trek thus celebrates a humanity defined by a unitary high culture untroubled by cultural difference. Yet the Enlightenment discourse of human progress, while promising universal and self-evidently virtuous values, always carries the cultural traces of its originators. Humanism, the rights of man, reason, freedom of speech, belief and movement are proclaimed as timeless and unimpeachable rights. In fact, they were the levers employed by the nascent classes of the Democratic and Industrial revolutions of the late eighteenth century; a way of challenging the hegemony of the landed classes, the Church and Monarchy and gaining a share in the political and wealth-creation processes. Thus the rhetoric of freedom of trade, association, right to be free of bonds of tradition, superstition and deference perfectly reflected the aspirations of French intellectuals, American plantation owners and British industrialists, who probably did believe that they were proclaiming self-evident and universal values grounded in an appeal to the beneficent properties of human reason and self-determination. Unfortunately, the majority would never have dreamed of admitting their wives, daughters, slaves and factory hands to a share in the same common humanity. The result is, however, that 'the term "human" is . . . conflated with a concept of Western Man' (Boyd, 1996: 101) which implicitly privileges the virtues of bourgeois, White, rational masculinity. *Star Trek*'s vision of exemplary humanity is thus founded on consensus and appeal to rational principles, but it is ultimately a vision of homogeneity, rather than radical diversity: 'His is the humanist drive towards "perfection" rather than the cyborg call for "optimization" . . . Data chooses to suppress his difference in order to reach toward the transcendent human ideal' (Boyd, 1996: 107).

This conformity effectively means, however, that any differences are effaced. Diversity must never threaten the smooth functioning of the ship. Only in the crew's personal quarters, in the décor and artefacts, in the peculiarities of hobbies and tastes, do we glimpse any degree of distinctiveness to do with

---

[13] All conducted with a clinical passion. Data's favourite detective, Sherlock Holmes, was of course famed for the rational and scientific approach to his enquiries.

ethnicity, race or gender. But officially, Federation culture, as Witwer notes, observes punctiliously the canon of high culture (Witwer 1995: 272–5). So the crew of the *Enterprise* entertain themselves in amateur dramatics, console themselves with Shakespeare and embellish formal occasions with chamber music (*Insurrection*, film, 1998). Yet is this not a strangely anachronistic world-view for a series so ostensibly futuristic? 'TNG actually presents culture as less problematic than it has been in recent years . . . The canon is intact again and cultural artifacts give a neutral, timeless, unproblematic representation of what it means to be human' (Boyd, 1996: 97).

Retrograde they may be, but Captain Picard is the embodiment of these values, and Data eagerly aspires to such pastimes as being the authentic hallmarks of full humanity. The study of Shakespeare is in many ways the litmus test of Data's aspirations. Under the influence of an intoxicant, Data misquotes from *The Merchant of Venice* in an encounter with a disbelieving Captain Picard:

Picard: Data, intoxication is a human condition. Your brain is different. It's not the same as . . .

Data: . . . we are more alike than unalike, my dear Captain. I have pores; humans have pores. I have fingerprints; humans have fingerprints. My chemical nutrients are like your blood: if you prick me, do I not . . . leak? ('The Naked Now', 1987)

Despite its comic overtones, Data's self-conscious evocation of Shylock's plea for acceptance sets the agenda for his portrayal throughout TNG.[14] Data's personhood seems to be forever conditional, a source of constant testing and ambition on his part and that of others. During another conversation about Shakespeare, Picard tells Data, 'You are here *to learn about* the human condition, and there is no better way of doing that than by embracing Shakespeare' ('The Defector', 1990, my emphasis). Picard's (rather patronizing) comments suggest that Data will always be an observer of the human condition, never a participant.

Zygmunt Bauman (1998) has characterized the crisis of just such a subject as epitomized in the figure of the parvenu. The exemplary figure of an ethic of humanistic self-improvement, the parvenu is characterized as one who migrates in search of success and self-validation according to modernity's unstinting standards. Within modernity, identity is a constant challenge, a project upon which the subject must expend energy, invest resources and spend time. In keeping with the logic of modernity, with its values of novelty, ambition, progress and hope, the self is bombarded by the relentless pressure of self-improvement. Hope and guilt characterize the modern subject, driven by the promise of success, but always falling short. In reading Bauman, I am reminded of Max Weber's characterization of the Calvinist entrepreneur in *The Protestant Ethic and the Spirit of Capitalism* (1902). A theological apprehension of a remorseless universe in which

[14] At a deeper level, Data may be understood as identifying himself with other generations of outcasts – Jews, African-Americans and other objects of racism (see Wilcox, 1996).

no final assurance of one's salvation could ever be attained precipitates the relentless urge towards maximization of effort in the pursuit of wealth creation, thus providing one of the preconditions for the emergence of capitalism.

The restlessness of the modern subject is due to the crushing weight of expectations of arrival and belonging. Quoting Hannah Arendt, Bauman speaks of the parvenu's fate in being 'denied the right to be themselves in anything and in any moment' (Bauman, 1998: 26). By definition, however, the demands of modernity are never satiated, and so hope is transformed to guilt – although the dream of self-realization must be sufficiently strong not to turn to despair. Parvenus must keep denying whatever they may once have been in the name of the endless fiction of modernity's ambitions. To win acceptance, parvenus must attempt to pass as real and deny the artifice of their performance; they have, even, to deny the palpability of their ambitions.[15] Like the monster, the parvenu both bolsters and destabilizes the condition from which it deviates, simultaneously perpetuating the endless round of self-invention and exposing its very fragility through its studied efforts. 'The parvenu's stay must be declared temporary, so that the stay of all the others may feel eternal' (Bauman, 1998: 24).

Repeatedly, Data learns about what it means to be human from his colleagues, who are perceived as the effortless and axiomatic experts. None of them seem interested in reciprocating, and in making a journey into the realm of 'cyborg citizenship' (Gray, 1997), or in considering how Data's artificial status renders any of their humanistic axioms problematic. In similar fashion to a monster of antiquity or the parvenu of modernity, Data shows forth both similarity and difference. He embodies simultaneously the value of Federation (that is, liberal Enlightenment) ideals of personhood through his tireless strivings to attain full humanity; yet his failure demonstrates his incapacity as machine ever fully to reach his goal. Data's position on the boundaries between human and almost-/ non-/in-human serves to reinforce and reiterate a particular state of affairs, rather than to invite a new journey into the realms of post/humanity.

Under closer scrutiny, therefore, TNG's core concept of an individualistic humanity serviced but not compromised by advanced technologies appears ill-prepared to cede to the immersive or incorporative powers of technoscience. The world created by Star Trek's progressive humanist narrative may turn out to be more problematic than appears at first glance; not so much a vision of utopia as 'an uneasy dream about human variety and human virtue' (Witwer, 1995: 271). For all its visions of technocratic futurism, TNG is highly equivocal about the advent of the post/human, tending to use its gallery of technologized humanity and humanoid technologies as boundary creatures against which the

---

[15] Data's odd chrome-like skin colour introduces racial undertones to his portrayal. His pallor is both reminiscent of Boris Karloff's make-up in James Whale's film production of *Frankenstein* and of the white-faced minstrels of the American South; an unreal whiteness that is perhaps intended to signal its artifice (Wilcox, 1996).

quintessentially human – in the enduring image of the rational, individualistic subject – is defined.

The prospect of threat to *Star Trek*'s liberal humanism becomes a source of horror in the form of another alien species whose hybrid organic–cybernetic constitution hints at post/human subjectivities that are the antithesis of Enlightenment individualism. Anxiety about this species, the Borg, takes place against the backdrop of enduring preoccupation in TNG with questions of identity. To see TNG as a comic genre, dedicated to the maintenance of identity in crisis, enables a clearer understanding of the gravity of the Borg's threat to this core ethos. The ultimate exemplars of this threat are those, like the Borg, whose complete and irrevocable integration of organism and machine has cost them their personhood.

Humanity under siege: the Borg

In their collective state, the Borg are utterly without mercy, driven by one will alone, the will to conquer. They are beyond redemption, beyond reason. (Captain Janeway, quoting Captain Picard, *Voyager*, 'Scorpion', 1997)

The fragility of selfhood, and its restoration, is a central concern of TNG (Richards, 1997: 64). Episodes frequently involve doubling or multiplication of selves. Thus Worf finds himself trapped in a proliferation of possible universes in 'Parallels' (1993); Data discovers a *doppelgänger* (Lore) in 'Datalore' (1989) and later 'Brothers' (1990) and 'Descent' (1993). In 'All Good Things' (1994) Picard shifts between three alternative selves. Attacks on identity are also launched through psychological torture or brainwashing, so that in 'The Best of Both Worlds' (1990), Picard is captured by the Borg, and in 'Frame of Mind' (1993) he is tortured. Troi is subjected to telepathic abuse in 'Man of the People' (1992). In 'Conundrum' (1992) and 'Naked Now' (1987) alien or external forces rob the crew of their usual identities. Other plots echo the danger of external encroachment upon the integrity of self, such as parasite possession ('Conspiracy', 1988), a prosthetic implant as instrument of torture ('Chain of Command', 1992) or the invasive and addictive potential of technology ('Hollow Pursuits', 1990; and 'The Game', 1991). Many plots therefore contain some kind of liminality, or a process of dislocation and crisis which requires the person to experience life outside normal routines or social contacts. This dislocated state is believed to generate a greater sense of purpose and integration; or it may ease a period of transition by enabling the individual to abandon certain roles and resume new ones. However, the pattern of 'severance and reintegration' (Richards, 1997: 94) is never left unresolved. In all these situations, restoration of character, of self, is achieved successfully by the end. 'Individuality persists throughout every surprise the universe can throw at the Enterprise [sic]' (Richards, 1997: 47). Episodes usually end in the unveiling of a mystery or the resolution of conflict. Characters may learn and develop as a result, and many episodes focus on characters making discoveries about themselves from which they emerge better persons. This is one aspect of the essentially comic character of TNG. Tension and conflict end in laughter and reconciliation:

In its pursuit of rounded character *Star Trek* is forever probing the absent spaces in the lives of its major characters. A sort of principle of negative space plays a very large role in imagining character in the series: the series explores not only who and what they are, but who and what they are not, and who and what they could have been. (Richards, 1997: 73)

Even in the twenty-fourth century, there are still enemies for the Federation, and the Borg emerge as the most convincing and evocative. The Borg are invincible, assimilating entire civilizations throughout the galaxy, which have their cultures 'downloaded' and thereby incorporated into the collective hive. The Borg are frightening because they pose a threat to the very continuation of human life itself. Assimilation is a euphemism for genocide. 'Borg' is a shortened version of 'cyborg'; and the depiction of the Borg shows them as organic bodies fully integrated with cybernetic prostheses. Their bodies are functional and broadly identical: black leather, pallid white skin (no racial identity). Prosthetics and cybernetic parts are, in the style of postmodern architecture, outside the skin. Beyond that, however, the Borg have assimilated individual identity into a collective consciousness. When Picard himself is captured by the Borg and assimilated into Locutus of Borg ('The Best of Both Worlds', 1990), therefore, it is as if a death blow has been aimed at the Federation's strategic and ideological core. Picard represents the finest values not just of the *Enterprise*, but of Starfleet and the Federation as a whole. He is a kind of twenty-fourth century *philosophe* – rational, cultured (and French). Passionate in his commitment to the humanistic ideals of progress, self-improvement and tolerance, he adheres conscientiously to the fundamental Starfleet doctrine of the 'Prime Directive,' which protects cultural self-determination for newly encountered civilizations.[16] His assimilation therefore exemplifies the terrible threat of the Borg, in the extinction of human identity and individuality at the hands of impersonal machinic political or technological systems. This would be, indeed, a fate worse than death:

Borg Voice: Strength is irrelevant. Resistance is futile. We wish to improve ourselves. We will add your biological and technological distinctiveness to our own; your culture will adapt to service ours.
Picard: Impossible! My culture is based on freedom and individuality.
Borg: Freedom is irrelevant: you must comply.
Picard: We would rather die! ('The Best of Both Worlds', 1990)

[16] Created for the OS, although more honoured in the breach than the observance, the Prime Directive counselled strict non-intervention in other, especially technologically less advanced cultures. Many critics believe this to have been veiled critique on the part of Roddenberry of US policy in South-East Asia, and of the Vietnam War in particular.

Picard's capture by the Borg thus throws into relief the utter contrast between the humanism of Starfleet and the totalitarian values of the Borg. The transformation of Picard, who embodies the Federation and therefore idealized humanity, into the Other betrays the fragility of the boundary. In his case, the transformation is something akin to a physical assault as instruments penetrate and probe, invading the integrity of the skin. Picard is not merely cyborged by his capture; he is robbed of his humanity and dismembered. As Locutus – a voice – he is no longer a complete person, but merely one functional part of a colonized body.

Eventually, however, the Borg are defeated by a strategy that highlights the contrast between creative individualism and the monoculture of the collective, and which stands for the essence of humanity's distinctiveness from the technologized collective: the superiority of human creative imagination. Knowing that firepower alone would not deter them, Picard's remaining crew realizes it must resort to its own innate powers of creative, even lateral, thinking. The Enterprise uses Locutus to transmit an instruction which sends the Borg collective to sleep. By seizing those qualities that, in the Star Trek world, distinguish humanity from machines, the Enterprise can – literally – outwit the powerful but unwieldy mentality of the Borg.

In later TNG encounters with the Borg, human creativity continues to be the antidote to the invasive and dehumanizing effects of technology. When a solitary Borg is captured and held prisoner on the Enterprise ('I, Borg', 1992), he identifies himself only as 'Three of Five' – a functional and impersonal title. As he is befriended by the crew, the Borg begins to respond as an individual; and to avoid having to refer to him constantly as 'you', crew members slip into calling him 'Hugh'. The act of naming marks an important transition, that of the acquisition of individuality, 'from abstract to concrete, from collective to individual, from alien to almost human' (Richards, 1997: 51).

The humanization of Hugh also sows the seeds of another ingenious ruse to defeat the Borg. A plot is hatched to return Hugh to the collective, not primed as a timebomb, but as a more subtle weapon. Hugh's emergent personhood will foster a similar process within the Borg collective, effecting a kind of 'velvet revolution' of rehumanization. As with the trick command to put the Borg to sleep, Hugh's emergent humanity – the essential qualities of individuality – will serve as the most potent weapon of all.[17] When driven to extremes, therefore, humanity's survival depends not on the extent to which it can use technologies or brute force, but on its ability to exploit enduring human qualities:

Picard: I wonder if the emperor Honorius, watching the Visigoths coming over the seventh hill, truly realized that the Roman Empire was about to fall? This is just

[17] Hugh's actions are later revealed as having played the Borg into Lore's hands. Their collective purpose weakened by individuality, they are susceptible to Lore's authoritarian manipulations ('Descent', 1993).

another page in history, isn't it? Will this be the end of our civilization? Turn the page . . .

Guinan: This isn't the end.

Picard: You say that with remarkable assuredness.

Guinan: With experience! When the Borg destroyed my world, my people were scattered throughout the universe. We survived, as will humanity survive. As long as there's a handful of you to keep the spirit alive, you will prevail – even if it takes a millennium! ('The Best of Both Worlds', 1990)

Such an emphasis on the enduring self in opposition to the technologized Other reveals the ambivalence at the heart of TNG's attitude to technology and science. TNG values technology (as a kind of reified, abstract good) as an aid to self-actualization (S.F. Collins, 1996: 141). It is, as many critics (and fans) of the series have observed, often used as a device to cut the Gordian knot of intractable plotting with a piece of technical wizardry. At its best it facilitates comic restoration; occasionally the source of temporarary confusion or danger, it will eventually enable its human masters to effect an ingenious solution. However, it never compromises or inhibits individual freedom – unless, of course, it falls into the wrong hands.

The Borg's technologies, however, represent a much more aggressive scientific telos, threatening to undermine human integrity, invading like a virus, infiltrating, seizing control and gradually mutating the host to support their existence. The Borg collective offers no route by which humans might enhance or augment their powers, but only a threat to those most precious, because most taken-for-granted, qualities, of individualism, self-determination and physical integrity. The Federation's anxieties about the Borg therefore rest in the cyborg body as anathema to the humanist self because it is compromised, hybrid, profane – monstrous. The Borg/cyborg body, in its mutability and hideous visibility, is effectively a feminized body, and thus anathema to a set of values which still privileges mind, reason and the transcendence of contingency. Even in their costuming, the Borg serve to shock, with their grotesque prosthetic limbs, external circuitry and tubing – like the monsters of late antiquity and early modernity, their repulsive nature derives from their visible malformation. The 'displacement' of organs and limbs, classically held to define a monster (Braidotti, 1994a: 78), plus their ghostly pallor and blank faces – signalling the absence of free will – are all visible signs of monstrous aberration. Technology in the guise of the Borg is voracious, practically primal, in its unstoppable urge to possess and engulf. It is a recapitulation of the 'Frankensteinian' anxieties of science out of control, and the defences of the individual, rational subject are perceived as no match for the totalitarian leviathan.

Despite the ubiquity of technologies, and the emergence of artificial beings, therefore, a clear distinction between humanity and machine remains central to the humanistic vision of TNG. The fundamental telos of the series is to protect the integrity of human distinctiveness, premised upon clearly demarcated boundaries between humans and others. Technologies are the benevolent servants of humanity, but will never be allowed to become an invasive or dominant force that may compromise notions of identity or rob humanity of its individuality. Given the degree to which many of their twenty-first century ancestors are already assimilating technological elements into their organic bodies, however, twenty-fourth century humanity displays a remarkably retrogressive resistance to the encroachment of digital, prosthetic and medical technologies on the integrity of the organic body: 'Most of the technologies of a starship are fully external to the human body. Its warp engines provide transportation, its phaser banks provide defence, its sensors provide information . . . Throughout the series the great divide between man [sic] and machine remains fully intact' (Richards, 1997: 40).

Science is, therefore, servant and tool, designed to enhance human powers, to extend its range of exploration, to broach frontiers; but never to confound or reconfigure the humanist self. Its purpose is to mimimize human vulnerability. This is a vision of 'the high development of technology creating a space for a rarefied humanness' (Rosenthal, 1991: 92). Anything which demonstrates a more 'porous' subjectivity, in which technologies may help to constitute identity, is unthinkable and anathema to *Star Trek*'s core values. The principles of Enlightenment humanism have their value, serving to protect individuals from torture, in ensuring democratic freedoms, enabling civilizations to envisage social and political orders based on dignity and respect for life. But the 'paranoia' ('Locutus', 1996) of TNG in the face of what resembles the post/human reflects confusion between humanity's interaction with technologies and its assimilation by technologies. As I argued earlier, the pristine subject of modernity is a representation of what it means to be human constructed to suit a particular historical, economic and political context but unsuited to a world in which human beings co-evolve with their artefacts and their environment. 'In other words, *Star Trek* says that we should be worried about the loss of a unified, discrete, autonomous body unit grounded in a self, and, at the same time, that we should champion subjectivity the way the crew embrace the formation of a discrete identity . . .' ('Locutus', 1996: 8).

Humanity restored: Seven of Nine

The saga of how one might reverse the terrible effects of Borg assimilation – the fate worse than death – has been a major preoccupation of the latest *Star Trek* series, *Voyager*. The character of the Borg drone, Seven of Nine, plays a similar role to Data, as the outsider whose presence throws into relief the axioms of human behaviour. In Seven's status as 'neither human nor Borg', human identity is portrayed less as an innately self-evident destiny than as an elusive and even baffling process of becoming. Seven's humanity develops within a community of unconditional acceptance but is nonetheless fraught and fragmentary. Seven must therefore journey through hitherto neglected dimensions of selfhood, such as psychology and spirituality.

This reflects an altogether more ambivalent attitude in *Voyager* towards many of the assumptions of earlier series. Even though they are separated by a mere decade, the second and fourth series of the *Star Trek* saga articulate values and concerns that are noticeably distinct. *TNG*, conceived and launched during Gene Roddenberry's lifetime, displays far more continuity with the assumptions of the original series, whereas *Deep Space Nine* and *Voyager* represent a sea change in crucial respects. While embodying *Star Trek*'s traditional values of liberalism and tolerance, it is also clear that *Voyager* inhabits a more complex universe in many ways, and Seven's re-entry into human society is no exception. Many of the traditional bearings and landmarks of earlier series are missing. Stranded in a distant portion of the galaxy by an alien force ('Caretaker', 1995), the ship is struggling to traverse the astronomical distance home. The loss of contact with Federation authority and mores, plus its need for self-reliance in a frequently hostile territory, place *Voyager* in a world in which there are few fixed points of reference, and few permanent values. Temporary alliances – even, at one stage, with the Borg – are justified if they manage to shave a few more light years off the voyage home. The geographical estrangement from the Federation diminishes the power of external authority; the novelty of situations encountered calls for a greater degree of strategic and pragmatic decision-making.

This is a world where the confidence of universal humanist precepts is beginning to disintegrate and alternative values must be exercised; and this shift from earlier axioms is especially apparent in the contrasting views towards religion between the series. In keeping with Gene Roddenberry's rationalist convictions, the first two series have little sympathy for organized religion, beyond portraying it as divisive and anachronistic; yet later series chart a mellowing of this antipathy (Barrett and Barrett, 2001).

Roddenberry was brought up a strict Baptist although by the end of his life had come to embrace a non-aligned theism. In an interview with the Jesuit Terrence Sweeney, Roddenberry declared that 'I think we intelligent beings on this planet are all a piece of God, are becoming God' (Alexander, 1996: 568). The idea of a supreme being or deity was equated with a cosmic consciousness as it transcended the barriers of selfishness, contingency and misunderstanding (Boeke, 1997: 52). However, while Roddenberry's spiritual convictions are articulated through his strong (if rather inchoate) belief in human perfectibility – 'I can see us moving toward that infinite oneness which is full of wisdom and peace and so on' (Alexander, 1996: 572) – any formal or institutional expression of religion was unacceptable. Spiritual or religious dimensions to characters and cultures were not regarded as consistent with the illuminating and emancipatory power of reason. Religious beliefs and institutions, however benevolent, are regarded as insufferable hindrances to human self-determination. This view is exemplified in the OS episode 'Who Mourns for Adonis?' (1967) where the Greek gods are revealed as interplanetary beings who keep their 'mortal' subjects in a state of ignorance. In the film *Star Trek V: The Final Frontier* (1989) a religious leader is portrayed as a manipulative demagogue, with a potentially destructive self-belief in his own status as divine vessel.

TNG continues the overwhelmingly secularist emphasis. In such episodes as 'Who Watches the Watchers?' (1989), 'Devil's Due' (1991), 'Rightful Heir' (1993), religion is depicted as harmful and illusory, a view consistent with the Enlightenment vision of the incompatibility of reason and faith. In 'Who Watches the Watchers?' a bungled anthropological observation of the Mintakans, a pre-technological race, has resulted in their mistaking Federation personnel and technologies for divine visitation. Picard invokes the Prime Directive to rebut suggestions that he collude with the resulting religious revival in order to cover up the earlier breach of non-interference:

Picard: Millennia ago, [the Mintakans] abandoned their belief in the supernatural. Now you are asking me to *sabotage* that achievement; to send them back into the Dark Ages of superstition and ignorance and fear? No!

In the most recent series, *Voyager* and *Deep Space Nine*, religion is depicted as a more vibrant source of social renewal and cultural integrity (Frederickson, 1998). Religious belief and practice is predominantly a tool for the deepening of individual self-understanding, as in the practice of the spiritual exercises of his Native American heritage by *Voyager*'s First Officer, Commander Chakotay. As Seven of Nine, the recovering Borg in *Voyager*, learns more about human nature, she reflects upon her nascent humanity and whether it necessarily embraces 'a sense of the holy'. 'The Borg have assimilated many species with mythologies to explain such moments of clarity. I've always dismissed them as trivial. Perhaps I was wrong' (*Voyager*, 'One', 1998).

In *Deep Space Nine* religious belief and practice is acknowledged as an institutional and cultural force. Captain Sisko is required to assume a formal role as the incarnation of the deity of another culture, the Bajorans. As well as being obliged to explore his own spirituality, Sisko encounters differing theological and epistemological debates to do with the compatability of science with religion, and also with the political implications of fundamentalism ('Emissary', 1992; 'In the Hands of the Prophets', 1993; 'Rapture', 1997).

Arguably, the definitive vision remains firmly shaped by Roddenberry's legacy. The essential faith in scientific rationalism is moderated but never abandoned. Science rules over mysticism; technology is an unqualified good; the 'progress' of the human subject is a given (S.F. Collins, 1996: 141; Porter and McLaren, 1999). However, the relaxation of earlier secularism may be symptomatic of a greater latitude in other areas too, which will be further apparent as I trace some of the other changing articulations of what it means to be human in the most recent *Star Trek* series.

The *Voyager* world contrasts with that of TNG, which inhabited a reasonably secure universe, in which enemies were clearly identifiable, one's own duties readily apparent and inviolable, and in which the constants of human identity, stability and order were protected and restored by the end of each episode. *Voyager*'s world-view has attained a more compromised, if not tragic, tenor. Frequently, dilemmas are not resolved, or order is restored in such a way as to leave further questions unanswered. In the episode 'Tuvix' (1996), a transporter accident merges two leading characters, Tuvok and Neelix. In its inception, the plot employs a classic *Star Trek* device – the dissolution of individuality, occasioned by invasive technologies – to explore the contours of human personality; but a search for a solution is clouded by ambivalence as to the right thing to do. As the crew struggle to find a means of separation, the amalgam third character, who has been named 'Tuvix', begins to explore his own unique identity. He pleads for his own life on the basis of sentience, self-determination and personhood – exemplary *Star Trek* values. Yet the decision is made that he must submit to his own death in order to separate Neelix and Tuvok.

Tuvix fights for his life to the end. The ship's doctor refuses to carry out the procedure in violation of the Hippocratic oath, forcing Captain Janeway herself to administer a lethal injection. The episode ends as Janeway leaves sick bay, her face in heavy shadow. The audience is left in no doubt that the restoration of two familiar characters will occur at the expense of the death of a third with whom it has been encouraged to sympathize. This is not a world of easy solutions, but one in which moral action is frequently a question of choosing the least harmful option.

Into this context is introduced a new character. An alliance with the Borg to fight a common enemy ends abruptly, stranding a Borg drone, Seven of Nine, aboard the *Voyager*. It emerges that she was once human, assimilated as a child, and raised as Borg. Successive episodes explore Seven of Nine's gradual acclimatization to human society. This is no comedic restoration of willing selfhood, however. Seven of Nine is resistant to a transformation that involves the acquisition of what she regards as human imperfection. She surrenders her Borg identity reluctantly, seeing no apparent advantages in making a human preference:

Seven of Nine: You are erratic, conflicted, disorganized. Every decision is debated, every action questioned, every individual entitled to their own small opinion. You lack harmony, cohesion, greatness. It will be your undoing. ('Scorpion', 1997)

If Seven of Nine is to be restored to full humanity, then *Voyager* resists the temptation to make this a painless transformation. At the start, Seven is reluctant to surrender her Borg heritage, and it is clear that her past cannot simply be erased.[18] This is consistent with *Voyager*'s internal logic. Seven's odyssey towards humanity will be as complex and uncharted as the spatial and moral landscape of the Delta Quadrant.

Nevertheless, the traditional *Star Trek* values of imagination, emotion, altruism and tolerance still endure as definitive benchmarks of true humanity. Seven's human colleagues instinctively evoke the qualities of affectivity that, if Sherry Turkle is to be believed, are held to mark the fault-lines between humans and machines:

Janeway: Imagination, creativity . . . fantasy . . . Human progress – the human mind itself
          – couldn't exist without them. ('The Raven', 1997)

These are not only the possessions of self-actualized individuals, however, but are also virtues practised in the context of community. What humanizes Seven ultimately will not be her ability to emulate or even mimic essential human traits, but her willingness to become human by learning the value of *belonging*:

Janeway: We have something the Borg could never offer: friendship. ('Scorpion')
Tuvok: You were Borg. But you are human now. You are part of our crew. ('The Raven')

Thus, while the qualities of integrity, imagination and authenticity are still prized, *Voyager* displays a more communitarian bias. Everyone – even Janeway – needs the reassurance of their comrades in tackling the confusing, even fearful, implications of taking responsibility for their own actions in a world where ethical certitude is dissolving. Humanity is not an innate property, but a state towards which all must struggle; a set of characteristics to be acquired amid a plethora of values and role models affirmed by one's own deeds in the thick of uncertain circumstances. Human becoming is only attained in relation with others who also perhaps share the unmapped journey of self-creation.[19] Identity, as a self-reflexive project, as neither entirely of our own making nor imposed upon us, will only be found, retrieved or remade by the act of belonging. Seven's crisis of identity is met with the reassurance of companionship:

Seven: I am no longer Borg but the prospect of becoming human is . . . unsettling.
Janeway: You belong with us. ('Hope and Fear', 1998)

Star Trek and representations of the post/human

Data and Seven of Nine are two science-fictional characters especially exercised by questions of what it means to be human. Like monsters, their liminal position

---

[18] Despite her acclimatization to Voyager, Seven continues to identify strongly with the Borg ('Drone', 1998; 'Dark Frontier', 1998).

[19] One of Seven's most significant tutors in human affairs is himself an outsider whose own personhood is a construction. The holographic Emergency Medical Hologram – the ship's doctor – is himself a digitally generated artificial life-form (Okuda and Okuda, 1999: 132–3, 804).

in relation to a wider human community serves to reiterate, sometimes to foreclose, but also occasionally to subvert, its ontological hygiene. I have argued that the overall tenor of TNG is romantic and comic, focusing on the restoration of order to situations that threaten the integrity of character. It serves to confirm an ordered world of an 'eschatalogical return to harmony, the cosmic overcoming of evil, and the redemption of the elect' (Farley, 1990: 12). What TNG fears most is the breakdown of the individual humanist subject, either by technological encroachment, by alien possession or by other forms of fragmentation, doubling or substitution. TNG portrays Data as a character on the verge of humanity, whose aspirations to become more human – some comic, others poignant and full of pathos – illuminate important questions. Data's search for greater self-knowledge is not a journey into a potentially new definition of subjectivity, however, but a perpetuation of modernity's most highly valued precepts of human nature, namely progress, self-improvement and individualism. An imagined twenty-fourth century is remade in the image of mid-twentieth century values.

The comic mood of TNG is tempered by a more sombre tone in the later series, Voyager, where the sovereignty of Enlightenment secular humanism is muted. Plots revolve less around external threats invariably resolved than explorations of the inner dynamics of individual personalities and the Voyager collective. Sometimes members' actions are morally questionable; frequently characters continue to struggle with painful dilemmas. In classic literary terms, Voyager is closer to the tragic genre than the comic. Tragedy manifests itself in narrative structures that speak of the deep ambivalence and irresolution of life: '[T]ragic vision locates the possibility of suffering in the conditions of existence and in the fragility of human freedom. The very structures that make human existence possible make us subject to the destructive power of suffering' (Farley, 1990: 29). Classically, the tragic heroes/heroines are brought low by the very qualities that elevated them in the first place, and so in Voyager the crew may still encounter alien forces 'out there' but dramatic tension is as likely to be derived from individual struggles with inner demons. Just as Voyager is adrift in the far-off Delta Quadrant, so in the lives of its crew, courses must be set without the co-ordinates of moral absolutes or the promise of clear resolutions or happy endings. The guiding vision of Voyager lies not in an unimpeachable humanity secure within clear moral boundaries but in the quest to discern some degree of purpose to the universe amid the complexity and fragility of everyday experience.

CHAPTER SEVEN

'Nietzsche gets a modem'[1]:
transhumanism and the technological sublime

Humanity looks to me like a magnificent beginning but not the last word. (Freeman Dyson, quoted in Regis, 1990: 146)

In Chapter five, I examined how technoscientific representations had come to occupy an increasingly definitive role in shaping exemplary notions of post/humanity in a digital and biogenetic age. In Chapter six, I argued that visions of post/humanity in humanist science fiction remain committed to a generic but predominantly rational human nature, exemplified in the virtues of individual freedom and self-determination. Technologies enhance, but never compromise, these essential human qualities. In this chapter I examine further representations of the interface between humanism and technology, in the shape of various examples of futuristic optimism. The seemingly unlimited potential of techno-science not only for building a new world order but for facilitating a new breed of post/humanity forms a common thread throughout this chapter. Such representations of the post/human therefore herald a transition into a technoscientific utopia assisted by smart drugs, prosthetics, genetic modification and computer-assisted communication. This is a vision of post/humanity augmented by 'artificially enhanced evolution' (Terranova, 1996: 165), thereby vanquishing the entropic forces of physical or intellectual finitude, morbidity and mortality.

Encompassing the popular science predictions of mainstream commentators, such as Michio Kaku and Ray Kurzweil, through to debates about the democratic potential of cyberculture, to the more expansive vision of transhumanism and Extropianism, these visions of a digital, cybernetic and biotechnological age subscribe to the values of 'Perpetual Progress, Self-Transformation, Practical Optimism, Intelligent Technology, Open Society, Self-Direction, Rational Thinking' (More, 1998: 2). Such advocates of enhancement and perfectibility would regard themselves as the twenty-first century heirs to Enlightenment traditions

[1] Kroker and Weinstein, 1994.

of secular liberal humanism, in which humanity, having displaced the gods, achieves heights of wisdom and self-aggrandizement – the post/human as superhuman. Indeed, the title of my chapter reflects the opinion of some that transhumanism, at least, is the fulfilment of Friedrich Nietzsche's vision of the *Übermensch* (see p. 173); but I will argue that a truly Nietzschean sensibility would regard such expansionism as fatally flawed by its inability to shed the vestiges of a Comtean 'religion of humanity', and is endemically incapable of acknowledging the absolute dissolution of value, hope and meaning which for Nietzsche was the destiny of humanity after the death of God.

If transhumanism might be regarded as a typical narrative of the secular world-view of 'science as salvation' (Midgley, 1992), then another vision of the post/human assumes, perhaps, a more unexpected complexion in its characterization of religion and spirituality not as antipathetic to, but fuelling, technoscientific innovation. Yet in their different ways, what David Noble terms 'the Religion of Technology' and the values of transhumanism sanction the pursuit of technological liberation in the name of what is assumed to be an innate human desire for immortality and omnipotence. A representation of the post/human as the inevitable outworking of the 'technological sublime' conceals many value judgements, however, not least in its universalizing of a metaphysics of technoscience founded on longings for invulnerability, incorporeality and omniscience. Far from being the impersonal outworking of a religious instinct, the technotranscendentalists' vision of the post/human, like that of the transhumanists, is thoroughly political. Who gets to participate in the post/human future? Whose desires fuel the priorities of technoscience? These things are not in the lap of the gods, and they require renewed attention to the significance of the appeal to 'transcendence' – and to particular constructions of religion – for exemplary and normative understandings of what it will mean to be human in the twenty-first century.

Technocratic futurism

Man's [sic] essence survives the vicissitudes of the body, with a brain of expanded functionality, with more highly evolved feeling, with further developed empathy . . . The web of Internet will truly become a body politic, loneliness banished for all, while maintaining individuality, privacy. People will not fall in love because of their appearances, but will love for its own sake, as foreshadowed in the music of Beethoven, of Bach. (Manfred Clynes, in Gray, 1995b: 53)

Early futurologists such as Daniel Bell and Alvin Toffler promised the advent of a knowledge-based society in which technologies of production and resource management would be conducted according to rational principles (Robins and Webster, 1988). The resulting 'knowledge society' promised prosperity, efficiency and social equity wrought by intelligent methods and technologies facilitated by

155

enhanced gathering and processing of information. In Marshall McLuhan's vision of the global village, the information revolution would enable a kind of digital democracy. More recently, politicians in the UK and the US have referred to the 'information superhighway' as the key to economic prosperity, largely as a result of the greater competitiveness of industry brought about by a more skilled workforce (Gore, 1994; Nellist and Gilbert, 1999). This is a vision of a post-industrial society that solves its problems of wealth creation and distribution through advanced technologies; a futurology characterized by technologically driven abundance and democratization.

Other recent visions of the future continue this mood. Entitled 'Your Bionic Future', a recent edition of the popular science journal *Scientific American* considered various dimensions of the future impact of genetic, cybernetic and digital technologies, setting out a panorama of 'How technology will change the way you live in the next millennium' (Agnew, 1999; Brown, 1999; Zorpette, 1999). Articles on AI, cloning, genetic modification and virtual reality offered real-life versions of the probable lifestyle of Western societies. The articles reflect a preoccupation with designer babies, cosmetic and spare-part surgery, cyber-shopping, smart houses and genetically attuned pharmaceuticals, representing priorities primarily comprised of cosmetic enhancement, leisure or lifestyle. There is nothing about the future of work, travel or communication, and very little, apart from an article about head transplants, concerned with curative medical applications outside genetic therapies. The editors might argue that their emphasis was intentionally on the quality of life in the future, but the overwhelming impression is of luxury lifestyle options that enhance the lot of those with the necessary resources to afford them. This is a vision of gain without pain: 'A genetic vaccine will increase muscle mass – without exercise' (Zorpette, 1999: 27).

Other optimistic advocates of the future prospects of digital and biogenetic technologies, such as Ray Kurzweil and Michio Kaku, paint similarly expansionist pictures of the benefits to humanity, such as longer lives, smart homes and devices, cheap resources and global communications (Kurzweil, 1999a, 1999b; Kaku, 1998). Recent developments in cybernetic, biomedical and information technologies are modest, it is argued, compared with the exponential increases forecast between now and 2020.[2] In Michio Kaku's vision of 2020, the classic economic problem of *scarcity* of resources has been superseded by unlimited plenty:

By 2020, microprocessors will likely be as cheap and plentiful as scrap paper, scattered by the millions into the environment, allowing us to place intelligent systems everywhere.

[2] Thanks to the growth rate articulated in Moore's law, which asserts that computer power doubles every eighteen months (Kaku, 1998: 14). I suggest the probable correlation of something we might term 'Gates' law', which states that personal computers decrease in retail price (and physical size) by a similarly exponential rate.

This will change everything around us, including the nature of commerce, the wealth of nations, and the way we communicate, work, play, and live. This will give us smart homes, cars, TVs, clothes, jewelry, and money. We will speak to our appliances, and they will speak back. (Kaku, 1998: 14)

Augmented and perfected by the latest innovations in artificial intelligence, genetic modification, nanotechnology,[3] cryonics[4] and other technological enhancements of human body and mind, the human race (or those intelligent enough to recognize the age to come) will be released from the chains of poverty, finitude, disease and ignorance to ascend to a higher, better, more evolved post/human condition. Technology does more than undergird humanist principles: it promises to create a successor species. 'Humans have beaten evolution. We are creating intelligent entities in considerably less time than it took the evolutionary process that created us. Human intelligence – a product of evolution – has transcended it' (Kurzweil, 1999b: 60).

Similarly unequivocal articulations of technocratic futurism may be gleaned from other popular cultural sources of the 1990s. Magazines such as Mondo 2000, 21C and Wired thrived during the 1990s and were largely aimed at, and written by, those in information technology, users of Internet discussion groups and readers of cyberpunk science fiction, and articulated the pretensions of a subculture which believed itself to be at the cutting edge of human–machine interaction, ushering in a successor phase of 'posthumanist' evolution. Mondo 2000 combined 'a chemistry left over from the 1960s' (Sobchack, 1993: 569) with the individualism and ambition of the neo-conservative

---

[3] Microscopic machines that mimic organic processes. 'Nanoengineering' and 'nanotechnology' derive from 'nano' meaning one billionth of a metre, or five carbon atoms. Its principles were first mooted in a lecture by the physicist Richard Feynman in December 1959, entitled, 'There's Plenty of Room at the Bottom', in which he argued that there was nothing in the laws of nature to prevent matter being manipulated at the most basic level possible, that of individual atoms. Nanotechnology is founded on a model of building life 'from the bottom up', atom by atom by molecule. Feynman's vision involved the building of tiny machines that would then build increasingly smaller versions of themselves. Feynman saw nothing in the laws of physics to prevent such an outcome. In the mid-1970s, Eric Drexler, inspired by the discovery of DNA, began to explore the possibilities of building synthesized molecules: mini-robotic devices capable of manipulating other small particles such as molecules. Why not use such nanotechnology to rebuild cells when damaged or dying? Nanomachines would go into the affected area and reorganize the DNA of the cells to revitalize them, molecule by molecule (Kaku, 1998: 266–71; Regis, 1990: 109–43).

[4] Cryonics traces its origins to a book entitled The Prospects of Immortality by Robert C.W. Ettinger, published in 1964. In it, Ettinger proposed a method of deep-freezing living tissue in order to preserve it until such a time when cures or recuperative treatments might be available. Ettinger was confident that, in time, every known disease would find a cure; and that modern science would also discover ways of slowing or even halting the ageing process. This technique Ettinger termed 'biostasis', meaning the artificial preservation of tissue at low temperatures awaiting recuperative intervention, by conventional means or nanotechnological cell repair machines (Regis, 1990: 85–8).

1980s,[5] setting itself as the harbinger of what it called 'the New Edge'[6]: a rich array of virtual realities, bio/nano/cyber-technologies, personal computing, electronic communications and global marketing opportunities afforded by the digital revolution. In characteristically hyped-up prose, Mondo 2000 promised its readers that they were living at the cutting edge of a new era in human history, announcing itself in a mixture of evolutionary and apocalyptic terminology: 'The dawn of a new humanism. High-jacking technology for personal empowerment, fun and games. Flexing those synapses! Stoking those neuropeptides!' (Mondo 2000, issue 1, quoted in Sobchack, 1993: 572).

Mondo 2000 tended to concentrate on the specific union of the cybernetic and organic, emphasizing in particular the radical changes to human bodies and minds effected by all manner of electronic media, but some Mondo 2000 features have also entertained the possibility of accelerated evolution occurring as a result of biogenetics or recreational drugs (Sirius, 1992). It set its cap at the 'sophisticated, high-complexity, fast lane/real-time, intelligent, active and creative reality hacker' (Sobchack, 1993: 573) – or, perhaps, those who imagined that a subscription to Mondo 2000 could transform them into just such a person. But Mondo 2000's ethos was firmly one designed to allay the stereotype of computer users as solitary losers, boldly proclaiming cyberculture as sexy, subversive, hedonistic and powerful. Mondo 2000's version of future humanity was one of technologically enhanced and genetically mutated bodies, transported into new virtual environments, resulting in a successor race of technologically enhanced beings 'whose symbiosis with the machine will be total' (Terranova, 1996: 167).

Terranova argues that the expansive optimism of such literature may be one of the more 'excessive manifestations of the cybercultural spirit' (Terranova, 1996: 175) but is typical of an implicit consensus among 'wired' subcultures of the time. Similarly, in talking of an emergent epoch of 'postbiological' humanity, those who call themselves transhumanists envisage that technologies will overcome the problems of physical limitations (of strength and intelligence) and finitude (decay, disease and death) by means of implants, modifications or enhancements.

One group of transhumanists is known as the Extropians, their name encapsulating their quest to defy the entropy of human bodily deterioration, disease and ageing; for the transhumanist spirit of technological and evolutionary inevitability expels defeatism and negativity, qualities that have no place in the Extropian

---

[5] One of its gurus is Timothy Leary, one-time advocate of pharmaceutically derived alternative consciousness, now advising his readers to 'turn on, boot up, jack in' (Dery, 1992: 510). The lyricist for the West Coast cult rock band the Grateful Dead, John Perry Barlow, has also contributed to Mondo 2000.

[6] This is presumably a play on the 'New Age' which, like its electronic sibling, is an eclectic synthesis of sixties alternative culture and late capitalist self-help consumerism (Ross, 1991: 15–74). In spelling out the nature of the New Edge, however, R.U. Sirius also characterizes the contemporary novelty of cybercultures as akin to Columbus sailing to the edge of the known world (Sirius, 1992: 16).

world: 'Where others say *enough is enough*, we say *Forward! Upward! Outward!* . . . Rather than shrinking from future shock, Extropians continue to advance the wave of evolutionary progress' (More, 1998: 5).

In their values of individualism and self-improvement, and in their eschewal of external authority or interference, Extropians espouse a libertarian philosophy. 'We recognize the dangers of controlling others and so only try to improve the world through setting an example and by communicating ideas' (More, 1998: 4). Government, law, the inertia of entrenched expertise are all impositions of authoritarian control. More's 'Extropian Principles 3.0' cite Eric Drexler, Hans Moravec, Richard Dawkins, Friedrich Hayek and Karl Popper, indicative of More's neoliberal, rationalist, secular humanist credentials.

The statements of Max More are the most accomplished articulation of the implicit values of technocratic futurism, in delivering a secular narrative of salvation through technology. Transhumanism is descended from the unalloyed faith in the primacy of the Enlightenment subject – rational, autonomous, self-determining. Machinic evolution will complete the task of natural selection; abetted by the fruits of technoscience, transhumanism advances the licence for exponential self-improvement:

A posthuman is a human descendent who has been augmented to such a degree as to be no longer a human . . . As a posthuman, your mental and physical abilities would far surpass those of any unaugmented human. You would be smarter than any human genius and be able to remember things much more easily. Our body will not be susceptible to disease and it will not deteriorate with age, giving you indefinite youth and vigor. You may have a greatly expanded capacity to feel emotions and to experience pleasure and love and artistic beauty. You would not need to feel tired, bored, or irritated about petty things. (Bostrom, 1999: 8)

As Regis acknowledges, it is tempting to see the scientific ambitions of cryogenics, nanotechnology and the transmigration of intelligence as fringe science, and the typical exponent of transhumanism as a practitioner of the bizarre and the hubristic, like a latter day Prometheus and Frankenstein, daring to cheat the gods and, doubtless, paying the price. Yet, as Regis comments, transhumanism is not dependent on supernaturally engineered visions, but objectives pursued through 'mainstream' science, albeit, to borrow from Regis, 'Science slightly over the edge': 'They'd take science to the point where the human species could step right up to the edge, look across the dividing line separating the Human from the Transhuman . . . and then cross over' (Regis, 1990: 278).

Whereas the hermeticists of antiquity evoked the spiritual world for aid, transhumanism's ambitions are the extrapolation of legitimate science, 'bringing science to the point where it would finally have caught up with what it was that human beings had always craved, which is to say, immortality and transcendence' (Regis, 1990: 278). Regis's analysis is consistent with my earlier critique of technocratic futurism and transhumanism. The fundamental conviction uniting

159

all such technoscientific dreams is that of the unlimited potential of human intelligence, lifespan and physical faculties. Such expansive proponents of scientific potential eschew ideas of there being limits to growth or inbuilt lifespans for human individuals. Mere biological finitude is no obstacle. In that respect, such models of technoscientific endeavour have inherited the values of eighteenth- and nineteenth-century humanism, not only in its technophilic embrace of the prospects of scientific innovation, but in its vision of humanity freed of the constraints of superstition, ignorance and fear and liberated to pursue a brilliant destiny. 'Like humanists, transhumanists favor reason, progress, and values centered on our well being [sic] rather than on an external religious authority. Transhumanists take humanism further by challenging human limits by means of science and technology combined with critical and creative thinking' (More, 1998: 1).

Such a vision of a post/human era is therefore in many respects a cybernetic version of social Darwinism, anticipating a future meritocracy founded upon the survival of the fittest, represented by the intellectual and psychological superiority of postbiological humanity. Paradoxically, however, while claiming to 'transcend' nature through technology, transhumanism also tends to defer to what it sees as the iron laws of natural selection. Like sociobiology, which appropriates 'cultural' behaviours to describe 'natural' processes (such that human traits such as 'selfishness' become projected on to the destiny of genes), so too humanity's deployment of technology becomes its means of transcending nature yet simultaneously remains in thrall to its logic of survival and adaptation. The result is a confusion of anthropocentric triumphalism and evolutionary determinism: 'Evolution's grandest creation – human intelligence – is providing the means for the next stage of evolution, which is technology' (Kurzweil, 1999a: 35).

THE POLITICS OF CYBERSPACE

Cyberspace is a place where bodies aren't supposed to matter, but many women discover that they do matter. (Balsamo, 1996: 150)

In a transhumanist world, the decentralized nature of the Internet is celebrated as facilitating a perfect future of democratic participation. As a virtual environment, cyberspace represents a populist and dynamic realm, free of centralized or bureaucratic control, in which cultural and social constraints dissolve. Howard Rheingold refers to cyberspace as a latter-day *agora*, a democratization of public space in which knowledge is power (Rheingold, 1991). Others clearly regard computer-mediated communications as effecting a kind of samizdat politics:

Digital technology tends towards democratizing communications. Ever-increasing numbers of human beings can send information and content directly to each other without the intervention of capital, a publisher, or an editor – presuming access to a modem and a computer ... From a historical perspective, while acknowledging vast numbers of information have-nots, we are still rushing headlong towards a new sort of human being – a creature with a 'voice' in the world. (Sirius, 1995: 3)

The anonymity and disembodiment of electronic communication is held to undermine formerly binding categories of race, gender, bodily ability and class. The physical distance and anonymity of the Internet grants cybernauts the option to experiment with new identities, and to sustain multiple personae in a variety of virtual media (Turkle, 1995). However, a note of caution is sounded by David Holmes, who speaks of the 'paradoxical' nature of cyberspace, of its capacity to shrink geographical distances and deliver the means for truly global communication while simultaneously reducing the scope of social interaction from the collective to the atomistic. The Internet and World-Wide Web exhibit contradictory features of 'connectivity and segregation' (Holmes, 1997b: 38), polarising interactions into those between the privatized self and the generalized global, with few intermediary social institutions. A sense of global solidarity, especially in the form of special interest groups, may be induced, but the bonds of everyday social integration do not receive the reinforcement of physical proximity (Pascoe and Locke, 2000).

While some commentators insist on the potential of computer-mediated technologies to create new, virtual communities, and to make Marshall McLuhan's vision of the 'global village' a present reality, therefore, others have queried how such a fluid and essentially ephemeral medium can sustain the social bonds and collective will necessary to establish any lasting political institutions (Nguyen and Alexander, 1996; Holmes, 1997a: 14–19). The conditions for Internet communities becoming 'genuine' communities may be undermined by their commitment to values of pluralism and decentralization. This results in a kind of 'flattening' of understandings of participation. Their transience and plasticity – the core of their spontaneity – may militate against more enduring and robust political forms.

GENDER AND POWER IN CYBERSPACE

When I explain virtual reality to the uninitiated, they just don't get it. But they warm immediately to the idea of virtual sex . . . I think lust motivates technology . . . The first personal robots, let's face it, are not going to bring people drinks. (Michael Saenz, cited in Springer, 1996: 82)

Similarly, feminist critiques of cyberculture suggest that the dynamics of gender and power mean that reality falls far short of a technocratic utopia. In the 'real' world, there is a preponderance of men as computer users, from academic and commercial computer science, to consumers of computer games and users of the Internet. Reasons given for this have ranged from the lack of encouragement to girls at school, to an exclusive culture in universities and business to the inhibiting power of sexual harassment (Spender, 1995: 177–8). The anonymity of electronic communication leaves women vulnerable to 'flame' mail or being bounced out of Usenet discussion groups by men attacking or belittling them (Cherny and Weise, 1996). Women logging into discussion groups or Multi-User Domains (MUDs)[7] report men bombarding them with explicitly sexual comments, requests for dates or sexual favours. Women participants in fantasy games in MUDs frequently find pornographic or violent scenarios directed towards them (Spender, 1995: 193–223). Research suggests that men who adopt a female persona also experience the same kind of treatment. Describing such sexual harassment and associated electronic intimidation as 'endemic', Dale Spender argues that such behaviour represents a critical threat to any utopian vision of the digital age:

Current practices are not about a few testosteroid males flexing their macho muscles and doing no real harm. Sexual harassment is becoming the modus operandi of the new world; the medium of communication for the generation of wealth and the support of power. It is the means by which some males are conquering and claiming the new territory as their own. This spells disaster for the hopes of an egalitarian ethos in this anarchic new world. (Spender, 1995: 210)

Spender advances some diverse analyses of the root of the problem. One, an inherently male pathology which simply transposes objectionable tendencies towards sexism, pornography and machismo into cyberspace; a second, the possibility that greater regulation will protect women and children from harassment; a third, that women-only spaces and special training – a kind of virtual self-defence course, helping to make women electronically 'streetwise' – will empower women to gain more space and confidence for themselves. The third proposal gains most sustained approval on Spender's part, representing a collective resolve on the part of women users of electronic mail, Websites and virtual domains not to be intimidated (Spender, 1995: 227–47).

Spender's strategies may address matters of access, but others adopt a more radical analysis, arguing that the very metaphysics of cyberspace is gendered. Those who deploy psychoanalytic (Freudian and neo-Lacanian) perspectives argue that cyberspace is but an extension of enduring and innate human erotic

---

[7] So-called Multi-User Domains are text-based or text- and graphic-based interactive domains in cyberspace in which a variety of interactions, games, role-plays are conducted. Some of the more sophisticated versions, such as LambaMOO, involve thousands of people (Dibbell, 1998; Turkle, 1995: 11). Participants log on and assign themselves a character; as they gain in expertise, they modify the programme and effectively alter the ambience of the game for others. One evident attraction for participants, and a matter of note for observers, is the licence for inventing new persona through the fantasy characters. The multiple, role-playing self is given carte blanche to roam freely in the virtual world: and clearly, the possibilities for creating multiple 'selves' – virtual and real – has implications for an anatomy of the posthuman text-based or text- and graphic-based interactive domains in cyberspace in which a variety of interactions, games, role-plays are conducted (Kelly, 1994; Turkle, 1995; Bromberg, 1996; Dibbell, 1998).

drives (Heim, 1993a, 1993b); but it is interesting to note the implicit gendered assumptions lurking therein. Michael Heim regards the flight into virtual reality as the continuation of a desire – an 'ontological continuity' (1993a: 63) – to project one's physical self beyond its own limitations; to touch others, to reproduce or to create new instruments and forms of life. For Heim, 'the creation of computerized entities taps into the most powerful of our psychobiological urges' (1993a: 66). Sandy Stone elaborates, characterizing an engagement with cyberspace as 'protean' (Stone, 1993: 108), an acting out of deep-seated drives 'to cross the human/machine boundary, to penetrate and merge . . .' (1993: 108).

In describing such a desire as 'cyborg envy', Stone may well be ironically noting the gendered logic of such a model of technophilia. Heim also criticizes the anti-materialist and somatophobic tendencies lurking within technotranscendence and technochantment – the drive to escape bodies, death and finitude. The yearnings for the godlike omniscience of cyberspace are themselves fuelled by a desire rooted, paradoxically, in the very self-limitation that it seeks to overcome. Born of eros that requires mystery and alterity, the ordered codes of Platonic information kill its prerequisite, desire. The loss of physical presence, of material situatedness, entails a diminishment of moral sensibility, because the ethical imperative of the unknowable autonomous other no longer pertains: 'The ideal of the simultaneous all-at-once-ness of computerized information access undermines any world that is worth knowing. The fleshly world is worth knowing for its distances and for its hidden horizons' (Heim, 1993a: 80).

One strand emphasizes the nature of cyberspace as virgin uncharted territory, into which the human creative urge must penetrate (Benedikt, 1993), while the other evokes a metaphor of cyberspace as matrix or web in order to reclaim virtual technologies as sites of feminist empowerment (Plant, 1995). These interventions remind us that cyberspace is political space; it may be a site of post/ human ontology, but the metaphors used to delineate its possibilities also exemplify significant assumptions about gender and power.

'NOT EVERYONE MAY FIND A PLANETARY CULTURE TO THEIR TASTE . . .'

It is a matter of debate, however, whether the digital and biotechnological revolutions have afforded opportunities to usher in egalitarian and inclusive forms of political agency or whether the biotechnological and digital age will simply enrich the privileged few at the expense of impoverished nations (Penley and Ross, 1991a). The dream of transhumanism depends for its fulfilment on the ability to have access to the appropriate resources. Much technological innovation is a Western commodity. While most of those with the resources and access to enjoy advanced technologies would stress the pleasurable qualities of prosthetic, digital or biomedical enhancement, it does not follow that it is necessarily a universally or unconditionally liberative prospect, immune from material inequalities. To privileged First-World citizens, the digital and biotechnological

163

developments bring with them an expansion of selfhood beyond the limits imposed by finite bodies and minds. To those unable to participate, however, it means further exclusion, compounded by the possibility that due to global-ization, the wealth of Western cyborgs rests on the cheap labour of their Third-World sweatshop fellows (Brasher, 1996: 817). Accelerating global digitalization does not simply revolutionize personal communications: it is part of a global economic revolution – a transition towards a post-industrial age – which re-quires fewer and fewer workers. Those advantaged by these structural changes stand to benefit, to become a new 'techno-elite' (Sirius, 1995: 3); but at the expense of millions of a cyber underclass. For all the democratic potential of cyberspace, those already 'information-rich', either through greater expertise or superior resources, will benefit at others' expense.

Despite their iconoclasm and self-styled radicalism, transhumanist principles have little to say about human evolution as involving the eradication of poverty, disease and discrimination. In its advocacy of libertarian anti-authoritarianism, transhumanism sidesteps collective or structural solutions to political and eco-nomic ills. The vision is for those assertive, wealthy and well placed enough to realize their vision unhindered. Similarly, for Michio Kaku, the rush towards a global culture 'unparalleled in human history, tearing down petty, parochial interests . . .' (1998: 19) is unquestioningly axiomatic. These 'parochial inter-ests' are, for him, conservative impediments to the onward march of scientific progress; but what if such interests involved, for example, the defence of rain-forests, endangered languages or minority cultures? Kaku's assumption that global communications will efface cultural diversity is entirely plausible, but he chooses not to consider the ethical or political implications. He accepts as inevitable the loss of 90 per cent of living human languages, and the triumph of multinational corporate interests, while conceding that 'not everyone may find a planetary culture to their taste' (Kaku, 1998: 337). The projection of human prospects, extrapolated to 2020, 2050 and beyond, contain no consideration of global inequalities; but this is not simply because explicit questions of equity and distribution are obscured. It is also a function of a model of the nature of technologies themselves that refuses to regard the choices behind the design and dissemination of particular technologies as embodying certain economic inter-ests or cultural values. 'What human beings are and will become is decided in the shape of our tools no less than the action of statesmen [sic] and political movements. The design of technology is thus an ontological decision fraught with political consequences' (Feenberg, 1991: 2).

Other projections into a technoscientific future reveal different priorities. The United Nations Human Development Programme (UNHDP) has identified the inequalities inherent in the information revolution (Jolly, 1999; Denny, 1999). It records that 93.3 per cent of all the users of the internet originate from the richest 20 per cent of the world's nations; 6.5 per cent of those online come from the next 60 per cent, leaving only 0.2 per cent of all internet access to the poorest 20 per cent of the world's population. As the report observes, 'New information and communications technologies are driving globalization – but polarizing the world into the connected and the isolated' (Jolly, 1999: 5).

In contrast to the prospect of smart houses, microprocessors and gene therapies, the UNHDP has set very different targets for global development by 2015: goals to combat illiteracy, reduce child mortality, eradicate poverty and to promote primary health care (Jolly, 1999). While the readers of Scientific American can look forward to smart jewelry programmed with software enabling them to 'order their favorite brew from a robotic coffee machine' (Brown, 1999: 72), the technologically rudimentary prospect of furnishing every man, woman and child on the planet with clean water goes unaddressed. The desires of a minority of the world's population (the 20 per cent already online) fuel the priorities of technoscientific innovation; but in the context of transhumanism's representations of the post/human, it is clear that they leave the interests of a significant minority unrepresented.

Technochantment

[O]ur technologies embody our aspirations, as well as our accomplishments. (Pitt, 1995: 5)

While exponents of transhumanism regard themselves as the heirs of the secular Enlightenment, Ed Regis may be nearer the mark when he speaks of the principles undergirding Extropianism as not so much scientific as metaphysical (Regis, 1994: 2). '"We see this need for transcendence deeply built into humanity," said Max More . . . It seems to be something inherent in us that we want to move beyond what we see as our limits' (Regis, 1994: 7). The philosophies and practices of transhumanism exhibit a will for *transcendence* of the flesh as an innate and universal trait, a drive to overcome physical and material reality and strive towards omnipotence, omniscience and immortality. That, rather than sustainability, justice, or even wealth creation, constitutes the metaphysics of transhumanism. According to Noble, however, technoscience has always betrayed its spiritual aspirations, even in the digital and biotechnological age:

Although today's technologists, in their sober pursuit of utility, power, and profit, seem to set society's standard for rationality, they are driven also by distant dreams, spiritual yearnings for supernatural redemption. However dazzling and daunting their display of worldly wisdom, their true inspiration lies elsewhere, in an *enduring, other-worldly* quest for transcendence and salvation. (Noble, 1999: 3, my emphasis)

Science and religion are often portrayed as antipathetic, but recent challenges to this interpretation argue that it has only predominated over the past 200 years,

165

and is really only 'secularist polemic and ideology' (Noble, 1999: 4). Scientific endeavour has been legitimated by theology since the Middle Ages (Wertheim, 1997, 1999) and a deep-rooted religious impulse is held to inform technological innovation (Noble, 1999). It is not, says Noble, that Western culture, seeing the sterility of science, is returning to religion to compensate. There has never been a fundamental contradiction between the two disciplines; and Noble contends that scientists are as full of messianic and apocalyptic expansiveness about the transcendent potential of scientific innovation as ever. The current fascination with new technologies as capable of synthesizing new worlds, what Jane Bennett characterizes as 'technochantment' (1997: 17), is akin to more ancient modes of spirituality, for both seek to 'transcend' the contingencies of the profane and material world in search of the more enduring realms of heavenly perfection. Signs of a preoccupation with omnipotence and transcendence are still apparent, argues Noble, in the psychological investment humanity places in technologies, not merely as items of convenience and utility but as instruments of 'deliverance' (Noble, 1999: 6).

Artificial Intelligence advocates wax eloquent about the possibilities of machine-based immortality and resurrection, and their disciples, the architects of virtual reality and cyberspace, exult in their expectation of God-like omnipresence and disembodied perfection. Genetic engineers imagine themselves divinely inspired participants in a new creation. All of these technological pioneers harbor deep-seated beliefs which are variations upon familiar religious themes. (Noble, 1999: 5)

The apotheosis of the twentieth-century quest for divinity through science, betraying a thoroughly dualistic concept of the world, comes in Noble's discussion of a speech by the geneticist, French Anderson, one of the leading practitioners of gene therapy in the US. It is worth quoting Noble's account at length:

In November 1991, Anderson delivered a talk . . . entitled 'Can We Alter Our Humanness by Genetic Engineering?' Anderson recounted the triumph of his DNA-deficiency experiment and attempted to evaluate the ultimate consequences of manipulating 'the very core of our being.' Was there the danger that we would somehow thereby distort or diminish our humanness, our defining essence? He rhetorically philosophized the traits that characterize us as human. Disclosing his own belief in 'a supernatural Being' and 'a resurrected soul', he concluded that there was no cause for concern, because humanness resides not in the body at all but in the 'soul' – '. . . that non-quantifiable, spiritual part of us that makes us uniquely human.' 'If what is uniquely important about humanness is not defined by the physical hardware of our body', Anderson argued, 'then since we can only alter the physical hardware, we cannot alter that which is uniquely human by genetic engineering . . . We cannot alter our soul by genetic engineering.' However much we might manipulate the physical, material components of our living beings, therefore, our essence survives untouched – 'the uniquely human, the soul, the image of God in man.' (Noble, 1999: 199)

Warming to his theme, Anderson defends research into the human genome on the grounds that it will never compromise the eternal 'essence' of humanity, namely our immortal souls.

Noble's thesis is echoed elsewhere, as other writers offer further expositions of science as manifestation of a drive to 'transcendence'. Michael Lieb's extremely original reading of the book of Ezekiel characterizes the writer's celestial visions of a heavenly chariot as part of an impulse towards what he terms 'the technologization of the ineffable' (1998: 3). In contrast to a materialist perspective in which humanity transforms the natural environment into commodities to which cultural value is attributed, Lieb reverses this relationship, characterizing the human creative impulse as the drive to render tangible an invisible reality, and claiming that this drives scientific and technological endeavour. Lieb offers a case study in the connections between a supposed 'spiritual' dimension of human experience and a technological imperative. The vision of God inspires its guardians to express that apprehension of reality in technologies: 'For them Ezekiel's *visio Dei* represents the wellspring of the impulse to fashion a technology out of the ineffable, the inexpressible, the unknowable' (Lieb, 1998: 3). The vision of God contained in the book of Ezekiel in the Hebrew Bible has informed subsequent accounts of fantastic technologies, including nuclear apocalypse, right through to contemporary UFO sightings: 'In this world, the emergence of technology confirms the ability of humankind to gain control of its environment and to overwhelm its enemies. This ability is grounded in the "will to power" that derives its impetus (as well as its putative legitimacy) from the ultimate source of power, God himself' (Lieb, 1998: 1).

The original vision of the writer of Ezekiel, of a heavenly chariot (or *merkabah*) is, according to the prophecies that bear his name, a manifestation of the Holy One. Its appearance establishes it as a physical object, yet possessing a numinous quality that unambiguously speaks of its divine origins. This brilliance is the same power harnessed by successive generations of human technologists who seek to implement similar mighty power through artefacts and tools. Technological innovation thus functions 'as the occasion through which Ezekiel's vision becomes a source of discovery, a means of knowing, of perceiving the true nature of the ineffable and how the ineffable operates in the world as we know it . . .' (Lieb, 1998: 4).

This visionary imperative seems to take precedence over other possible motives for technological innovation, such as greater efficiency of production, new sources of fuel or raw materials, the impulse to transform a hostile environment (or, indeed, to exploit a fertile one). In that respect, Lieb assumes that the 'religious' impulse is both compellingly universal *and* that it always manifests itself in a flight from the material, immanent world. For although Lieb is talking about a religious vision made flesh, the assumption with which he begins is that the divine image is immaterial to begin with. Lieb thereby articulates another example of the making real of divine pretensions as the wellspring for technologies. The fabricated world is derived from an impulse to transform the substance of (in Ezekiel's case) visionary dreams into physical commodities, a representation of technoscientific activity as worship.

Technopaganism

Erik Davis also sees a continuity between virtual technologies and religious sensibilities. In his entertaining discussion of 'technopaganism' he argues that in their strong convictions that the digital world is animated and enchanted, that information technology is the continuation of hidden codifications of wisdom known only to initiates, and in their recovery of neopagan rituals to celebrate not only the cycles of nature but the wonders of technological invention, such high-tech adepts are the direct descendents of the hermetic magi (Davis, 1993, 1995, 1999). Analogies can certainly be drawn between the hermetic imagination and aspects of digital and biogenetic technologies, especially those involving virtual and information technologies. I alluded to the hermetic tradition in Chapter four, and speculated on the possible cross-fertilization between gentile magi of the Renaissance and Kabbalistic Judaism. While hermetism was primarily a practical art, encompassing astrology, numerology, herbalism and magic, it also had a clearly identifiable set of philosophical convictions. In particular, Hermetism bears the stamp of Gnosticism, which held that the physical universe is vibrant with divine energies, whose forces could be appropriated by humanity. Beyond the visible world lurked a realm of metaphysical powers. Access to them, and thereby to the essence of creation, was through knowledge; not pure reason, but *nous*, akin to intuition. Thus the *Hermetica*[8] tells of the soul seeking salvation through mystical enlightenment as a means of attaining an approximation of the celestial intelligence which animated the cosmos. In order to achieve enlightenment and to experience the unveiling of divine truth, the adept must undertake a journey of the mind through the levels of the cosmos, assisted by the various practical arts that held the key to the mysteries of the universe. 'Therefore unless you make yourself equal to God, you cannot understand God: for the like is not intelligible save to the like. Make yourself grow to a greatness, beyond measure, by a bound free yourself from the body; raise yourself above all time, become Eternity; then you will understand God' (*Corpus Hermeticum XI*, quoted in Yates, 1979: 32).

Information and memory was all: 'ritual names, spells, and astrological correspondences; numerological techniques; ciphers, signs, and sigils; lists of herbs, metals, incense, and talismanic imagery' (Davis, 1993: 588). While neopaganism stresses the enchanted character of nature, technopaganism follows hermetism

[8] See Chapter four.

in regarding the technological world (and especially computer-mediated environments such as cyberspace) as redolent with hidden wisdom, enchanted spirits and magical patterns. The quasi-Gnostic themes inherent within Western techno-science thus engender a symbolic system of scientific enquiry as unveiling the hidden fragments of a deeper, transcendent realm which holds us enchanted by its magical prospects.

However, Frances Yates's classic study of hermetism argues that two strands of *gnōsis* informed the imagination of the hermetists, and can be seen as exercising their influence. One, which she terms 'pessimistic' or 'dualist' *gnōsis*, regards the material world as 'heavily impregnated with the fatal influence of the stars' necessitating 'an ascetic way of life which avoids as much as possible all contact with matter, until the lightened soul rises up through the sphere of the planets, casting off their evil influences as it ascends, to its true home in the immaterial divine world' (Yates, 1979: 22). The 'optimistic' strand sees material reality as infused with divine power, which the adept must evoke by means of secret and esoteric wisdom (1979: 22–3). The material world is thus both the source of all evil and the vehicle of heavenly transport.

Clearly, the science that has produced computer-generated communications is radically different from that of the alchemy, numerology, cryptography and other arts of the Gnostic metaphysicians.[9] The important connection lies in the significance of *information* as constituting the true essence of the universe in which there is an authentic world of pure form, the present material world being but a dim reflection. Similarly, the idea of DNA as pure code – and genes, as the bearers of that information, constituting the motivating force of life itself, possessing an autonomous and self-directing *telos*, echoes the Gnostic understanding of a cosmos driven by inner, eternal wisdom. 'Information, impatterned and wild, is the very context of life as well as the simulacra of essence, whether it is called consciousness, personality, individuality, or a unique cognitive system' (Gray and Mentor, 1995: 454).

CYBERSPACE AS SACRED SPACE

[C]yberspace is a metaphysical laboratory, a tool for examining our very sense of reality. (Heim, 1993a: 59)

Mindful of the possibility of cyberspace constituting a new kind of space, a dimension of pure information, some commentators speculate on cyberspace constituting a new kind of sacred space, 'a vast and sublime realm' (Davis, 1993: 586). 'The elsewhere of cyberspace is a place of salvation and transcendence.

[9] Among whom is discussed the Elizabethan metaphysician John Dee (see Chapter two). Note also the similarities between the Kabbalah and the Gnostic emphasis on the epistemological power of letters, numbers, codes and automata (Chapter four).

This vision of the new Jerusalem very clearly expresses the utopian aspirations in the virtual reality project' (Robins, 1995: 147).

In his lyrical evocation of the numinous beauty of cyberspace, Michael Benedikt suggests that virtual reality is but the latest manifestation of an innate human need to inhabit fictional worlds, an expression of a latent yearning for alternative, 'mythic planes' of existence (1993: 6). Benedikt speaks of cyberspace as 'the realm of pure information' (1993: 3); digital technologies, similarly, possess a spiritual energy redolent with a 'millennial spark' (Davis, 1993: 587). Technologies offer the opportunity to construct the celestial habitats that previously have only existed in the imagination; not so much the mastery as the transcendence of physical environments where at last, the creation of ethereal space enables the higher virtues to flourish as a digital utopia. Indeed, Benedikt deploys the Biblical language of the 'Heavenly City' of Revelation, which, in all its 'weightlessness, radiance, numerological complexity . . . utter cleanliness, transcendence of nature and of crude beginnings' (Benedikt, 1993: 15) fulfils human longings for enlightenment, perfectibility and order. The physical sensory world is but a reflection of a purer, Platonic realm. Matter becomes energy as physical entities are transformed into the pure forms of disembodied intelligence and code. The artificial language of binary code, divorced from the everyday conventions of material communication, can aspire to the lofty perfectibility of absolute form. For some, this presents a means of escaping this mortal coil and ascending into a brilliant celestial realm. Cyberspace becomes effectively a portal into another world; cyber cowboys (in William Gibson's terms) are 'riders on the chariot of the internet' (Lieb, 1998: 69). Online, bodies are abandoned in the 'bodiless exultation' of cyberspace (Gibson, 1984: 12), in a drive to overcome the particularity and frailty of embodiment which impels humanity to seek and assimilate into more ideal, more impermeable forms.

Cyberspace is said to be characterized by ambiguity, transition and indeterminacy (Tomas, 1993). Because it represents a disruption of 'normal' space and time, cyberspace is believed to constitute a portal into a sacred realm, offering transformations in time, space and consciousness (Stenger, 1993: 54–5). The disruption of temporal and physical conventions of time and space potentially offers transport into omnipresence and omniscience where 'every document is available, every recording playable', creating a playground, a library without limits (Benedikt, 1993: 2), affording access into a group mind never before experienced. It seems to be driven by a desire to 'become (like) gods', by striving towards semi-divine status and knowledge, either by contemplation or manipulation of the physical world as a means of gaining mastery and power over the world.

Cyberspace is indeterminate, too, in that it suspends 'normal' conventions of body, space, time and place. It is a world of sense without mediation, and representation without materiality. In the future, according to Rheingold, virtual reality could be used to induce altered forms of consciousness as in the past, cultures have used intoxicants, ritual or other extreme mental states: 'someday, in some way, people will use cyberspace to get out of their minds as well as out of their bodies' (Rheingold, 1991: 356).

Such rhetoric makes a particular connection between desire, disembodiment and spirituality. Michael Heim characterizes as primal the fascination with technology in its ability to excite fundamental desires and awakenings, whether they be spiritual or more prosaic. Our attraction to technologies, like moths to brilliant flames – 'the fiery objects of dream and longing' (1993a: 61) – is certainly aesthetic, maybe even erotic in nature. We seek the intimacy of connection, the consummation of desire, and computers are erotically charged, although Heim does not make clear whether the computer is the means or the end to the fulfilment of desire. Does the medium in some way construct the desire it feeds, or is it more the eternal outworking of innate and unchanging drives? 'Eros is a drive to extend our finite being, to prolong something of our physical selves beyond our mortal existence' (Heim, 1993a: 63). It seems as if this erotic drive for connection is driven, at least in part, by a fear of fleshly limitation, a limitation that may be both physiological and existential.

Part of the 'transcendent' nature of cyberspace rests in its disembodied detachment from physicality. All this reflects a deep-rooted world-view in which the physical sensory world is but a reflection of a purer, ideal realm of perfect form. For some, the advent of new information technologies represents the realization of this model. 'At the computer interface, the spirit migrates from the body to a world of total representation. Information and images float through the Platonic mind without a grounding in bodily experience' (Heim, 1993a: 75). Physical information can be transformed into the pure forms of disembodied intelligence and data; human ontology is digitalized.

It is certainly possible to detect religious language in the enthusiasms of exponents of digital and genetic technologies, as we have seen; but is it accurate to claim, as Noble does, that the longings for transhuman aggrandizement and technoscientific expansionism are the – 'enduring, other-worldly' – *inevitable* outworkings of an innate spiritual quest for transcendence of embodied finitude? 'Masked by a secular vocabulary and now largely unconscious, the old religious themes nevertheless continued subtly to inform Western projects and perceptions' (Noble, 1999: 104). Noble's use of 'unconscious' is intriguing here. Is it intended to mean 'implicit', 'secularized' or simply requiring no further explanation? Despite Noble's claims for the convergence of science and religion, a definitive connection remains unproven. There is a difference between saying that 'technology is the outworking of essentially and inescapably religious impulses', and 'religion and scientific rationality can, despite all the prophecies of the Enlightenment, coexist'. The former view posits some clear causal link between science and religion, but the latter merely states that one of the core

171

precepts of modernity – namely its secularism – may have to be re-evaluated. The fact that science has not extinguished religion, as many Enlightenment thinkers assumed, is no guarantee that science is the manifestation of an innate transhistorical spiritual instinct. Could such a scientific impulse not be articulated in some other way – perhaps in terms of psychological *cathexis*, for example – if it had not been manifested as a specifically religious world-view?

Grace Jantzen's exploration of the 'necrophilic' tendencies of Western religion continues David Noble's characterization of the scientific enterprise (Jantzen, 1998), although she challenges his assumption that the religious impulse behind technoscience is innate or universal. In its logic of salvation, in its valorization of immortality and fear of death, Christianity implicitly constructs a model of God as omniscient, disembodied and immutable:

The emphasis on the omnipotent, detached 'God out there' is ... not unrelated to the ideal of neutral, detached reasoning, which is in turn part of the fantasy of an ungendered, non-embodied rational mind, separate from all else, living towards a world beyond, expressive, in short of the imaginary of death. (Jantzen, 1998: 209)

However, Jantzen argues that this is but one historically contingent construction of divinity, one that acts as the ideological underpinning of a culture of death, domination and indifference. Rather than abandoning the religious world-view, we must attempt to reconfigure it, for even in our 'secular' age, religion remains an influential source of the prominent cultural symbolic. Jantzen argues for a reconfiguration of the fundamental dynamics of what transcendence, immanence and divinity might mean, founded upon renewed understandings of natality and flourishing as embodied and holistic. 'Divinity in the face of natals is a horizon of becoming, a process of divinity ever new, just as natality is the possibility of new beginnings' (Jantzen, 1998: 254). This holds out a number of possibilities for thinking about human engagement with technologies and with humanity's own creative potential (especially to create and enhance life and to transform our material surroundings with beauty and utility), and of continuing to consider how that engagement with the material might be an avenue into 'transcendence' or divinity. In the next two chapters, I want to advance an understanding of transcendence that synthesizes the material and symbolic without retreating from the world's messiness and contingency, an evocation of transcendence born more of 'translation' than 'purification'.

While I concur with Jantzen's project of reconfiguring the religious symbolic in order to dismantle the equation of religion and 'transcendence', it is also imperative to consider the economic and political dimensions. Ideas alone cannot be changed without addressing the material interests that inform them. It seems to me that far more than dualistic religion is feeding Western technoscience.

The pursuit of monetary gain is a considerable factor in determining the direction of scientific and technological change. There are huge profits to be made in the programmes associated with the Human Genome Project, not least by the private biotechnological corporations who are funding the research, patenting the 'genes' and marketing the applications (Hubbard and Wald, 1997). I suspect, therefore, that religion is less the sole underwriter of science *per se* than one of many ideological bolsters of a voracious consumer capitalism that encourages particular patterns of relationship and engagement with technology, 'nature' and social order. In particular, it is important to address the coexistence of the urged-for transcendence – a surrender of materialism the better to attain quasi-divinity – with the constant stimulation of consumer desires.

Nietzsche's modem

Kroker and Weinstein's reference to Friedrich Nietzsche as 'the patron saint of the hyper-texted body . . .' (Kroker and Weinstein, 1994: 2) occurs in the context of their paean to the technological sublime of virtuality, 'the perfect evolutionary successor to twentieth-century flesh' (1994: 1). Given his popular association with ideas of the 'overman' (*Übermensch*), who transcends Judaeo-Christian values and harnesses the 'will to power' in the pursuit of a courageous vision of self-actualization, Nietzsche may seem the perfect prophet of a libertarian, apocalyptic transhumanism. 'Nietzsche's got a modem, and he is already writing the last pages of *The Will to Power* as *The Will to Virtuality*' (Kroker and Weinstein, 1994: 2). This is ostensibly an open invitation for Extropians, transhumanists and technocrats everywhere to cast aside the outmoded constraints of ethics, altruism and humanism in favour of a technologically realized super-humanism, or an 'end-of-the-century liberation theology as Transformer robot' (Dery, 1992: 521). A closer reading, however, suggests that Nietzsche would have abhorred what he might have regarded as an excessive and uncritical transcendentalism.

Certainly Nietzsche shared the principles of secular humanism voiced by those such as Max More, in the latter's characterization of transhumanism as seeking 'the continuation and acceleration of the evolution of intelligent life beyond its currently human form and limits by means of science and technology, guided by life-promoting principles and values, while avoiding religion and dogma' (More, 1998: 6). More and Nietzsche stand in the tradition of Ludwig Feuerbach, inveighing against traditional theism. Feuerbach saw the exposure of the fiction of God as heralding the dawn of a new age in which humanity, acknowledged as the true authors of their world, could achieve emancipation, whereas Nietzsche, at the other end of the age of reason, saw no such promise of redemption. The death of God is necessary for the shattering of all other-worldly appeals to morality and truth, but with God's dethronement the entire edifice of morality, meaning and purpose crashes into ruins. Nietzsche saw this collapse as such a radical, iconoclastic event as to place the very survival of Western civilization in question.

Nietzsche regarded such an attrition as essential if humanity were to attain a new maturity. He conceived of the solution as a reworking of the Christian myth of sacrifice and resurrection, in a narrative of 'eternal return', repudiating temporal teleology in favour of a reborn figure. Nietzsche parodied the Christian narrative of the sacrificial suffering of Christ, using the figure of the ancient sage Zarathustra as a new saviour, who is reborn into innocence as the Übermensch. Having done away with the creator of the universe (God), the Übermensch must become the creator of his (sic) own reality. However, while humanity is now free (via the exercise of the will to power) to realize and effect its own destiny, Nietzsche was reluctant to reinscribe ultimate values or appeals to 'transcendent' visions. After the repudiation of Christianity – what he termed 'Platonism for the people' – all metaphysical systems are to be renounced. Dogma cannot be reinvented. Only a courageous fortitude endures, sufficient to 'light the lamps' in the darkness. This is a vision far removed from the values of transhumanism, which merely secularizes without abolishing Christian narratives of transcendence and redemption, offering a late twentieth-century version of Auguste Comte's 'religion of humanity' in which the imperatives of progress, unity and reason were to be serviced by a secular cult. In its aspirations towards the logic of immortality, invulnerability and omniscience, transhumanism exposes its vestigial craving for a perfect transcendent world 'Apart, Beyond, Outside, Above (Ingraffia, 1995: 92) the messy contingencies of this one. We might expect Nietzsche to hold transhumanism culpable for reiterating rather than transvaluing (purging, renouncing) a pathological dependence on metaphysics. His thought is more properly an indictment than an endorsement of the transhumanist endeavour.

Futuristic fantasies of how technoscience will create a prosperous world are politically charged. Exemplary and normative visions of what it means to be human fuel transhumanist ambitions, but they may leave some people unrepresented altogether. It is clear that further examination is required into the way in which a particular construction of 'religion' has been co-opted into the transhumanist project in order to naturalize its predilections for omnipotence, invulnerability and immortality. Advocates and critics alike of techno-transcendence assume that 'religion' and 'transcendence' can be equated; and that 'transcendence' is synonymous with Nietzsche's 'Apart, Beyond, Outside, Above' (Ingraffia, 1995: 92). However, the symbol of transcendence must itself be placed in a cultural, political and gendered context. In its representation as the apotheosis of human engagement with technology, it is important to exercise a *genealogical* reading of appeals to 'transcendence' (Carrette and King, 1998: 139–40). In respect of Nietzsche's genealogical thinking – which, of course, was influential upon

Foucault – it would be better understood as an attempt to call all value systems to account for their implicit idolatries, a question, ultimately, about what, in its representations of the post/human, Western culture has chosen to elevate as its objects of worship.

CHAPTER EIGHT

The end of the 'human'?

Technology, meant to extend our organs and our senses or even to support our fantasms of immortality and transcendence, seems to threaten what we wanted to preserve by destroying us as the subjects we thought ourselves to be when we took refuge in technological projects and dreams . . . Do we, by internalizing technology, lose ourselves as the 'subjects' of our culture? (Schwab, 1987: 81)

Chapters six and seven explored how humanistic science fiction and transhumanism both address the relationships and boundaries between humans and machines, the natural and the artificial, while strenuously maintaining the integrity of the modern Western humanist subject at the heart of their broadly technophilic visions. By contrast, in this chapter I consider representations of the post/human that entertain a variety of possibilities concerning the obsolescence, evolution or dissolution of human uniqueness. One of the earliest cinematic representations of technological themes, Fritz Lang's Metropolis (1927), captured contemporary fears of the loss of individuality in the era of mass production and central planning. The metaphorical representation of humans-as-machines has a longer history, however, and is frequently invoked within an evolutionary model of post/human development. Nevertheless, the language of humans-as-machines and the narrative of post/human evolution are best understood as representational devices that metaphorically delineate the normatively human in the face of challenges to ontological hygiene.

It will also become clear that much of the anxiety accompanying the invasive effects of technoscience focuses on the dismemberment or effacement of the body. Whether the body is discarded, retained or mutated in virtual media, it is clear that one of the effects of cyberspace is to render taken-for-granted concepts of embodiment problematic. With that goes a questioning of the a priori moral status of embodied presence as constituting what it means to be human, capable of articulating inviolable criteria for notions of subjectivity, participation and community.

However, the 'end of the human' means something different to those who eschew humanistic nostalgia; and some representations, such as those influenced by cyberpunk and postmodern perspectives, appear to be striving for models of post/humanity that move beyond the discourse of humanism. In contemporary genres such as cyberpunk and the postmodern performance art of Stelarc, the ontological hygiene between non-human nature, humans and machines is tested to its limits. Yet the consequences are to regard the boundaries of post/humanity as rhetorical and constructed, and effectively to move towards a more contingent and non-essentialist model of what it means to be human. It offers the possibility to portray the post/human as something that is always already mediated through, and constituted by, its environment and artefacts. Just as poststructuralism's 'death of the subject' attracts mixed reactions, however, so the promise of a proliferation of post/humanities without the illusion of an originary or exemplary norm is both exhilarating and perplexing. Can narratives of hope and obligation, and the vitality of political and ethical discourse, survive after humanism? Where might we go from there? And who would 'we' be?

Dehumanization and dystopia

Metropolis, premiered in Berlin in 1927, is a visually stunning landmark in cinematic science fiction. A deeply flawed work, it nevertheless established certain key conventions of representation for twentieth-century popular culture in its vision of the sinister and alienating effects of technology and of the futuristic city as place of wonder and oppression. The science-fiction motifs of automata, futuristic cities and oppressive machines furnish Lang with a vocabulary of representation with which to explore issues of economic and social relations and the encroachments of technology. It also offers insights into the influence of religion and gender in the formation of its representations of the post/human.

Despite its visual magnificence and innovative production techniques, there is much about the film, especially the plot, which is unsatisfactory. Set in the year 2000, Metropolis tells the story of a futuristic city, ruled by the autocratic industrialist, Joh Fredersen. Following an encounter with the saintly Maria, a workers' leader, Fredersen's son Freder resolves to alleviate the workers' monotonous and dehumanized existence. However, Fredersen senior conspires with a scientist, Rotwang, to duplicate Maria as a robot in order to discredit her. The plan backfires and the workers run amok, destroying the machines and flooding their underground city. After a rooftop chase, Rotwang falls to his death, Freder and the real Maria are reunited and together persuade Joh and the workers to be reconciled.

Joh Fredersen's wish to subvert Maria's influence seems illogical given that she is preaching quietism and patience. Rotwang's motives are obscure; and the final scene of reconciliation between capital and labour has been criticized as a

177

simplistic, even reactionary resolution (Jenkins, 1981; Jensen, 1969). Some of these problems may be explained by the degree of editing that apparently occurred during the production process, leaving many plot-lines under-developed or contradictory.[1] Nevertheless, the film abounds in vivid imagery that memorably articulates the implications of urbanization and mechanization in the form of political, psychological and gendered dichotomies that run throughout. One way of understanding the struggles between the various manifestations of the human and almost-human in Metropolis is as a play of difference around contested accounts of masculinity, rationality and self-control. Gendered representations that implicitly equate normative human subjectivity with the masculine consign qualities coded as 'feminine' – religion, affect, women – into the realm of that which threatens to extinguish the privileged human subject at the heart of the narrative.

Metropolis opens with shots of city skyscrapers intercut with factory wheels and the face of a huge clock (Jensen, 1979: 19), evoking immediately a landscape of machines and human-machines. The urban–industrial landscape is organized around the discipline of timekeeping; and the city's wealth and futuristic splendour rests on the systematic bodily and temporal disciplining of a dehumanized 'underclass'. A caption announces, 'The day shift', and the assembled workers move to their cell-like stations as one, suggesting their subjection to the physical and chronological discipline inherent in factory production. One solution to the tension between capital and labour is to do away with the human workforce altogether, and Joh Fredersen conspires with the inventor Rotwang to create a robot[2] that will take the place of human labour. Having deprived humans of their individuality, the logic of Fredersen's plan is eventually to supplant humanity altogether. Rotwang's robot will supercede the workers as the perfect new labourer, tireless, efficient and compliant. Seeking a way of lessening the influence of the activist Maria, Joh then conceives of a scheme to have the robot assume the guise of Maria in order to undermine her leadership and the workers' confidence in her. Not only is the robot an agent of human redundancy, therefore, but also a vehicle of deception; a thief of human freedom twice-over.

Meanwhile, the film turns to the character of Freder, whereupon Lang begins to deploy intriguing if unsubtle religious associations (Jensen, 1969: 12; G. Ward, 1999). Freder lives a playboy's life in the 'Eternal Gardens' at the heights of the city until, one day, he catches sight of Maria in the company of some city children and is immediately infatuated. Her madonna-like innocence and simplicity contrasts with the wiles of Freder's usual paramours. Anxious to see

---

[1] According to Paul Jensen, some seven reels may have been cut from an original total of seventeen (Jensen, 1969: 7).

[2] See Chapter four, note 21.

Maria again, Freder descends from the heavens to the factory level beneath the city and assumes the guise of a worker. He is soon exhausted by the physical labour and its monotony, and leaning against the giant dial of the clocklike mechanism at which he has been working, he adopts a cruciform pose and cries out to an absent Joh, in an echo of Jesus' last words from the Cross, 'Father, Father – I did not know that ten hours can be torture!' (Jensen, 1979: 52). Religious iconography is interpolated again when Freder infiltrates Maria's meeting for the workers in the catacombs. She appears before them in what seems to be a shrine, decorated with crosses. As she preaches her message of forbearance, Maria evokes qualities attributed to the Virgin Mary, as a pure, compassionate intercessor. But in foretelling the advent of a 'Mediator', Maria is also reminiscent of John the Baptist, antecedent of a greater prophet yet to come. Freder experiences a growing conviction that it is his destiny to be the very Mediator of whom Maria speaks.

Maria's solution to the dehumanizing instrumentalism of the Metropolis regime is one of a unity of head, hands and heart in harmony. In her homily to the workers she tells the tale of the Tower of Babel as one of a fatal division between those who conceived it and those whose labour created it. As Maria says, 'Between the brain that plans and the hands that build there must be a mediator . . . It is the heart that must bring about an understanding between them' (Jensen, 1979: 60). The juxtaposition of hand, head and heart explores the conflicts inherent in the urbanized factory environment, in which the ascendancy of logic results in the routinization of the workforce, the crushing of the spirit and the disenchantment of the world. The cult of Maria stands for the repressed qualities of affectivity and spirit, their subterfuge and clandestine character signified by their being driven (literally) underground and associated with the feminine. Maria's call for mediation is for the spiritual side of humanity (coded as feminine) to bring some 'balance' into the technical–rational world. But this is a domestication rather than a reconstruction of Metropolis's industrial order, reflecting a conviction that the dehumanized urban–industrial dystopia can be tempered and reformed by an injection of human(e) values (Kracauer, 1979: 16).

The contrasting figures of human-Maria and her robot *Doppelgänger* cast them as representatives of virtuous and dangerous femininity respectively, exposing 'the exceedingly fragile boundaries between good and bad science, good and bad beliefs, good and bad machines, and good and bad women' (Jordanova, 1989: 117). Human-Maria is portrayed as the antithesis of the world of industry and factory, a symbol of family, nature, affectivity and spirituality. Maria humanizes a heartless world of competition and discipline by preaching patience. Standing apart from the productive activity of the city, she does not constitute a direct threat to the survival of Metropolis. Rather, like Engels' analysis of reproductive labour, the feminine and the domestic are identified as serving to absorb and redirect the tensions of class conflict (Barrett, 1980). Once human-Maria has been captured by Rotwang and Joh – champions of the city's guiding principles of corporate business, scientific reason and discipline – and the robot transformed into her likeness, the new robot-Maria poses a far more immediate threat to the realm of technological power. Her 'feminine' qualities change from the pacific to the destructive as she threatens to destroy her creators. Before being sent to the workers, the robot performs an erotic dance for a party of industrialists. This is a portrayal of Woman as sensual and uninhibited; corrosive of Metropolis's work ethic of deferred gratification in pursuit of profit. The sexualized temptress – not virgin/mother, but Jezebel or Salome – appears as a threat to male self-possession, both the sexual propriety of bourgeois manhood and the disciplined labour of the workers. Robot-Maria is the embodiment of the erosion of human freedom and will, at sexual, psychological as well as political levels. Freder himself is placed in jeopardy, as a vision of robot-Maria conversing with Joh sends him into a nervous collapse, endangering his attempts to subdue the uprising. Freder's hysterical episode is intercut with scenes of robot-Maria's sensual dance, accompanied by images of death, as the statues in the cathedral – including one of the Grim Reaper – seem to come to life. The excesses of the robot-Maria threaten to undermine Freder's mission in a way that human-Maria, meek forerunner of a greater Saviour, never did. It is robot-Maria, however, who correctly diagnoses the workers' plight as she incites them to rebellion. 'Who keeps the machines going?' she asks. 'Who are the slaves of the machines? Destroy the machines!' (Jensen, 1979: 94–5). Her actions result in the destruction of the city, to the extent of endangering the lives of the workers' families, who are rescued by Freder and human-Maria, now free. Yet to the end, the robot exercises a disruptive influence, putting life, home and human survival at risk. After Rotwang's duplicity is revealed, the mob attempts to burn the robot as a witch – another symbol of female malevolence.

The two Marias form a binary pairing, one that frames an androcentric norm. As contradictory presence, embodying purity and danger, women cannot act consistently or independently, but only as mediator or catalyst to patriarchal power. Although Metropolis speaks of the need to acknowledge and integrate qualities designated 'feminine', such values are still ancillary to the world of men. In the final scene, it is Freder, true to his self-appointed vocation, who effects the reconciliation. Maria is now no longer central to events, except, possibly, as the future mother of an enlightened Fredersen dynasty.

For all its futuristic imagery, Metropolis is deeply conservative. Its corporatist union of master and servant and the representations of women suggest a collectivist and romantic ethos not unlike that of twentieth-century fascism. The vision of post/humanity within Metropolis suggests that machines and women are antipathetic to exemplary, androcentric, humanity. The feminine is simultaneously representative of uncorrupted virtue and untutored sensuality, neither of which are compatible with technological and civic functioning, which are founded upon values of reason and self-mastery. The female robot engenders only irrationality and chaos, hinting at the unreliability of women who, like machines, rob 'us' (coded as male, in a gendered unity that transcends the class division between labour and capital) of 'our' humanity (constituted as sexual control, reason, labour and freedom). Women are acceptable as guardians of domestic, subterranean (catacomb) and privatized values, but their ascent to the surface (the world of politics, technology and science) will only end in disaster. The heart will reconcile the alienated qualities of head and hand, although women cannot be integrated into the reconstituted whole, but must remain excluded 'others' within a redeemed civic society.

Humans as machines

[I]n polite company . . . it is not generally considered to be a good thing to allow oneself to be 'dehumanized'. (Law, 1991: 17)

For all its contradictions, the enduring image of Metropolis is of dehumanization. The figurative transformation of workers into automata and the literal mutation of the human-Maria into a robot are vivid representations of a mechanistic and disenchanted dystopia. Yet the metaphor of humans as machines has been a recurrent theme at least since the emergence of modern science during the Renaissance (Mazlish, 1993). Leonardo da Vinci characterized animals and humans as operating according to machine-like principles, enabling him to extrapolate examples of movement and design from 'nature' into his own sketches of artificial devices (Mazlish, 1993: 15). Mechanistic imagery informed theories of natural selection too: Linnaeus' taxonomies of plants and animals were attempts to prove an affinity between the development of organic life and the mechanistic laws of nature (Mazlish, 1993: 79–85). Medieval and early modern automata were primarily for entertainment but they also illustrated an emergent mechanistic model of the universe in which creation was conceived as a well-oiled piece of clockwork. Rene Descartes (1596–1650) appropriated this world-view and argued that animals, while alive, were not infused with divine spirit but animated by mechanistic forces (Noble, 1999: 143–8). God's immutability guaranteed the existence of an ordered universe that might be apprehended by mathematical laws. However, this had to be separated from subjective experience in order to be scientifically verifiable, so all sensory experience was evacuated from epistemology. Descartes argued, therefore, that reason distinguished human beings from non-human animals and gave them privileged knowledge of God.

The physician and philosophe Julien de la Mettrie (1709–51) continued the analogy that humans were machines with souls. 'The human body is a machine which winds itself up, the living image of perpetual motion' (La Mettrie, 1995: 205). Eschewing theological understandings of human nature, advancing a commitment to positivism and materialism, La Mettrie conceived of a continuity (albeit also a hierarchy) of species a century before Charles Darwin:

From animals to man, the transition is not violent, as good philosophers will admit. What was man before the invention of words and knowledge of tongues? An animal of his species, who, with much more instinct than the others, whose king he then considered himself to be, could not be distinguished from the ape and from the rest, just as the ape itself differs from the other animals; which means, by a face giving promise of more intelligence. (La Mettrie, 1995: 206)

At a theoretical and metaphorical level, however, the discipline of cybernetics, founded on a strong analogy between humans and machines, offers the most complete measure by which their respective intelligences are likened. In its emphasis on command, control and communication as the essence of sentient activity, cybernetics establishes a striking affinity between the organic and the machinic. Arguably, cybernetics represents the theoretical core of the corrosion of species boundaries in its conceptualization of 'the mechanization of the human and the vitalization of the machine' (Gray, Mentor and Figueroa-Sarriera, 1995: 5).

The principles of cybernetics[3] were prefigured in earlier theories, perhaps dating as early as Ktesibios' water-clock in the third century BCE (Kelly, 1994: 112). Self-regulated devices thrived under Islamic scientific scholarship in the early Middle Ages (Galison, 1994) and re-emerged in Europe in rudimentary form in various early industrial machines, such as the built-in safety mechanism in James Watt's steam engine, where a gauge (or 'governor') monitored and prevented the build-up of steam pressure (Kelly, 1994: 114–15). The pioneer of modern cybernetics was Norbert Wiener (1894–1964) a mathematical physicist at MIT who transferred into military research in 1941. While working on the design of a device to track and predict the flight of enemy bombers, he perfected the statistical theory from which a new science of communication engineering could be developed. The unpredictability of tracking hostile planes cost lives, and Wiener's ambition was to construct a reliable anti-aircraft system that would accurately calculate the flight patterns of enemy pilots, even allowing for their own responsive maneouvering to evade hostile ground-fire. In such a situation, Wiener argued, pilot and aircraft became symbiotically linked; and so it became strategically – and for Wiener, eventually, philosophically – expedient to consider human and machine as one entity: the 'servomechanism' (Galison, 1994: 233–40). Wiener increasingly came to see the anti-aircraft predictor as analogous to the behaviour of biological organisms: '[T]he physical functioning of the living individual and the operation of some of the newer communication machines are precisely parallel in their analogous attempts to control entropy through feedback' (Weiner, 1989: 26).

[3] From the Greek kybernētēs, meaning 'steersman' (Wiener, 1989: 15).

The philosophy underlying cybernetics lay in a notion of the universe as being constantly moving from a state of organization to 'entropy'.[4] Cybernetic systems are designed to ward off entropy, described as a 'measure of disorganization' (Wiener, 1989: 21), their adjustments compensating for any deterioration of function. Cybernetic mechanisms are highly interactive, capable of handling and responding to the demands of a continuous stream of information. They work on the basis of a continuous process of feedback between operation and environment – indeed, the degree of interaction renders the boundary between the two highly permeable – which allows the system to maintain a working equilibrium. By its very nature, however, the cybernetic machine is never at rest; any semblance of equilibrium is temporary. 'Life is an island here and now in a dying world. The process by which we living beings resist the general stream of corruption and decay is known as *homeostasis*' (Wiener, 1948: 95).

As the science of self-governing mechanisms, cybernetics attributes similarities of form to machine and human behaviour as resting in the propensity to process and adapt to external stimuli. Information is processed and translated into command and performance. The principles of vitality and negative entropy, therefore, rest on a dynamic of adjustment and reaction. The operation of this style of control and adjustment 'strives to hold back nature's tendency toward disorder by adjusting its parts to various purposive ends' (Wiener, 1989: 27). Wiener's cybernetics therefore represented the first sustained examples of systems of 'command and control' (Wiener, 1964: 2), enabling the design of fully autonomous and self-regulating machines. Wiener argued that the case for considering machines as alive lies in their self-regulating nature and their dependency upon the reception and processing of feedback and resistance to entropy.

Only by reverting to some notion of *essences* was it possible to distinguish between the liveliness of an organism and that of a machine. In principle, neither was more or less dead than the other. Life and death were no longer absolute conditions, but interactive tendencies and processes, both of which are at work in both automatic machines and organisms. Regardless of their scale, size, complexity, or material composition, things that work do so because they are both living and dying, organizing and disintegrating, growing and decaying, speeding up and slowing down. (Plant, 1997: 161, my emphasis)

Not only does Wiener conceive of a continuity between human 'intelligence' and that of self-regulating machines, but as Sadie Plant's use of the term 'liveliness' suggests, he also presents a model of human nature as dynamic and interactive

[4] David Porush argues that Wiener and an early collaborator, Claude Shannon, devised cybernetic functioning as a defence against Heisenberg's uncertainty principle in physics. If the presence of an observer introduced randomness, then the self-regulating logic of cybernetics would correct the disruption to perfect communication and control that such indeterminacy (portrayed as entropy) represented. Entropy is best understood as lack of perfection and precision than mere chaos. See Porush (1989: 375); Wiener (1989: 181).

rather than essentialist and monadic. Human intelligence is performative and, like much else of life itself, is characterized by the capability of organic and technological systems to monitor external stimuli and adjust operations accordingly. 'To live effectively is to live with adequate information. Thus, communication and control belong to the essence of man's inner life even as they belong to life in society' (Wiener, 1989: 18).

Much of Wiener's later work – after he had eschewed employment associated with any kind of military research – was of a philosophical nature, exploring not only the affinities between humans and machines but also the discontinuities (Wiener, 1964; 1989). One of the impulses behind *The Human Use of Human Beings* (first published in 1950) was to explore the ethical implications of mass production and mechanization, to ensure that new technologies would not transform workers into mere automata. Although Wiener was anxious to assert the distinctiveness of human experience, his influence on cybernetics has endured primarily in the founding premise that human and machinic intelligence is profoundly similar, although this affinity was properly to be conceived as functional and formal, rather than substantive.

CONTINUITY OR DISCONTINUITY?

Men are grown mechanical in head and heart, as well as head. (Thomas Carlyle, 1828, quoted in Mazlish, 1993: 65)

In *The Fourth Discontinuity* (1993) Bruce Mazlish advances the thesis that humanity in its history has traversed a number of 'discontinuities' that are linked by a common thread. Humanity and non-human nature are part of a continuum of existence, but this truth is systematically denied until scientific discoveries alert people to it. Thus Copernicus overcame the discontinuity between humanity and the cosmos, and by representing the universe as heliocentric he swept away the Biblical world-view and placed humanity within a different cosmological system. Similarly, Charles Darwin dissolved the hierarchy between humans and other species (another theologically derived orthodoxy) and required his contemporaries to admit their evolutionary affinity to apes and other higher mammals. Sigmund Freud challenged prevailing views regarding the primacy of the rational, conscious self by basing the nascent discipline of psychoanalysis on the profound continuity between adulthood subjectivity and infant sexual drives – albeit repressed and denied by the imperatives of acculturation. 'In this version of the three historic smashings of the ego, humans are placed on a continuous spectrum in relation to the universe, to the rest of the animal kingdom, and to themselves. They are no longer discontinuous with the world around them' (Mazlish, 1993: 4).

Mazlish argues that the so-called 'fourth discontinuity' awaiting humanity is that which distinguishes between humans and machines. Humans are sophisticated tool users, and their skill with tools has been one of the factors responsible for their evolution: '[A] human is that animal who breaks out of the animal kingdom by creating machines' (Mazlish, 1993: 213). As humanity took greater control of the processes of natural selection and evolution, 'along the way, tools began to turn into machines' (1993: 217). Gradually, therefore, the human race is being confronted with its affinity with alien or external phenomena. Before we know it, humans have – literally – turned into machines, mutating from organic into manufactured beings, under the influence of genetic engineering, cyborg implants and other advanced technologies.

However, the following extract suggests that Mazlish has confused two arguments in his discussion:

[T]he first involves human evolutionary development as being inextricably intertwined with tools and machines, and the second focuses on the development of one such machine, the computer, as inevitably making us take seriously the question whether the same conceptual schemes that apply to the 'thinking' machine apply equally to the brain. (Mazlish, 1993: 159–60)

Mazlish's first theme concerns human evolution, and argues for a continuity of tool use as defining human nature, an evolving capability driving the development of culture. The model of human as *Homo faber*, or tool-maker, is not novel. Sigmund Freud in *Civilization and its Discontents* argued that tools were merely extensions of human faculties: 'Man has, as it were, become a kind of prosthetic god. When he puts on all his auxiliary organs he is truly magnificent; but those organs have not grown on to him and they still give him much trouble at times' (Freud, 1995: 750). It is argued that contemporary digital technologies, such as virtual reality, are merely a continuation of human capacity for tool-making. What may seem unfamiliar and threatening is but an extension of perennial skills, serving not to undermine or expunge the human, but to enhance it, as tools have always done:

The continued western quest for making tools may . . . retrospectively be reinterpreted in relation to its culmination in virtual reality. From the club that extends and replaces the arm to virtual reality in cyberspace, technology has evolved to mime and to multiply, to multiplex and to improve upon the real. (Poster, 1996: 197)

This half of Mazlish's thesis thus advances the notion of a transcultural continuity to *Homo faber* as the essence of what it means to be human. It may be helpful, however, to be mindful of Donna Haraway's critique of representations of 'human nature' in twentieth-century primatology.[5] Whatever model of emergent humanity is posited, whether it be Man the Tool-Maker or Woman the Hunter-Gatherer, primatology deploys the almost-humans of prehistory to construct an evolutionary narrative that serves up 'human nature' in its own image

---

[5] See Chapter two.

and contemporary mores as the inevitable outcome of benevolent natural selection (Haraway, 1992b). Mazlish's implicit ideology of normative humanity begins to reveal itself by the end of the book, where he speculates about the motivation behind the characteristic human trait of tool-making. The ambition that distinguishes humanity is not simply the desire to transform the material environment for the purposes of comfort, survival or defence. It is the wish to transcend embodiment altogether, as driven by 'fear of death, loathing of the body, desire to be moral and free of error' (1993: 218), humanity picks up its tools and builds a world. The imperative of 'transcendence', interpreted as immortality, incorporeality and invincibility, discussed in the previous chapter, reappears here at the heart of Mazlish's speculations as to the origins of human tool use in prehistory.

The second aspect of Mazlish's thesis concerns the affinity between humans and machines, and the prediction that eventually humans will evolve into machines: '[H]uman biological evolution, now understood in cultural terms, forces upon humankind – us – the consciousness that tools and machines are inseparable from evolving human nature' (1993: 233). A discourse of 'transcendence' features here too, as Mazlish argues that 'part of being human is to seek to *escape one's humanity* by the creation of machines, which then reshape their creator' (Mazlish, 1993: 229, my emphasis). However, Mazlish literalizes what cybernetics treats as metaphor, namely the formal similarity between patterns of intelligent activity in nature and culture. He also co-opts another scientific discourse, that of evolution and natural selection, in order to telescope cybernetic affinity of function into a prediction of human obsolescence. But if cybernetics may be considered as a particular representation of affinities between humans and other forms of intelligence, so evolution too may just as well be conceived as a narrative about the superior adaptive skills of *Homo sapiens*, without taking it to be the literal prediction of the end of humanity altogether. The representation of the post/human as the inevitable and inexorable successor to the human, falls prey to what Keith Pearson terms 'cosmic evolutionism' (1997: 220). While metaphors of humans-as-machines are nothing new, as I have indicated, such a narrative has something of the self-fulfilling prophecy about it, because evolution, not human agency, then seemingly fuels technoscientific innovation. A metaphor has become a deterministic *telos*. It fails to consider the dialectical or emergent nature of human evolution, or to consider that 'biology' and 'nature' may adapt rather than dissolve in relation to technological change.

I am not convinced by Mazlish's narrative of the inevitable displacement of discontinuity by continuity. The continuity between humans and a series of 'others' – the cosmos, non-human nature, the unconscious and artefacts – is tempered by the various ways in which cultural representations might also serve to establish *discontinuity*. Leonardo and Linnaeus may have used motifs of continuity between nature and culture to inform their work (see above), but others emphasized human distinctiveness. René Descartes was concerned to reflect on ways in which, if at all, humans were different from animals. While he characterized the latter as machine-like, possessing no independent powers of free will or reason, the former were able to perfect a flawed and instinctual nature by means of their rationality. The passions of humanity's animalistic nature are compromised by embodied experience, but thought founded on reason clears all doubt and leads to sure knowledge.[6] The faculties which characterize genuine human nature are therefore, for Descartes, those by which knowledge may be verified without recourse to physical sensation.

Rather than being a narrative of impersonal and deterministic evolution, therefore, Mazlish's analysis might be more fruitfully interpreted as signalling a series of paradigms in which the boundaries between human and non-human are periodically redrawn. While Mazlish opts for a story about the supposed erosion of discontinuities, a more convincing account might be one of how scientific and popular representations of 'human nature' tell a succession of stories about humanity's relationships to its 'others' that are couched in terms of affinity *and* distinctiveness. Indeed, Mazlish peppers his discussion with metaphors of humanity such as 'perfectible machine' and 'neurotic animal', suggesting a range of (representational) strategies which articulate both continuity *and* discontinuity. Exclusive attention to the inevitable homology of humans and machines risks obscuring the shocks of technological transformation, and evades the reasons why people may insist on clinging to a particular discontinuity, a paradigm dismissed by Mazlish as an incomprehensible, almost irrational, resistance to inevitable and self-evident scientific progress.

The various representations of continuity and discontinuity may be paradigms of the human condition, but they cannot simply be assimilated into a linear, gradualist narrative of evolutionism. Mazlish's conception of humanity as inventive tool user may certainly have its attractions, but I prefer to return to the earlier metaphor of humanity as builder of worlds, material and symbolic (Chapter two). Humanity inhabits a universe of its own making in which the boundaries between human and almost-human are discursively as well as concretely reconfigured.

Virtual post/humanities

The relationship between body and self, already problematized by the 'cyborgization' of the physical body through prosthetic and genetic technologies, is also thrown into question by the spread of virtual forms of embodiment and electronically mediated communities. Human ontology is digitalized, yet if identity is reconstituted as pure data or information (Bey, 1998: 3) – as 'text' – then the

---

[6] Hence, *cogito ergo sum.*

corporeal 'texture' of embodiment becomes more problematic. The absorption of corporeal presence into a purely digital embodiment may well suggest the effacement of the body, such that participation in cyberspace is necessarily postcorporeal, 'a quintessentially cerebral experience, distilled from the material dross and distortion that is flesh itself' (Kirby, 1997: 135). This might be understood as the apotheosis of Cartesian and Enlightenment rationality; once liberated from the drag of embodiment, the self is master of pure knowledge. Certainly, much of the commentary on cyberspace assumes that bodies are, in Marvin Minsky's terms, simply 'meat' to be abandoned in order to allow 'pure' consciousness itself to roam free (Sheehan and Sosna, 1991). Yet if persons have no fleshly substance in cyberspace, then this raises the question of whether it is still appropriate or meaningful to link traditional ideas of identity, freedom, agency or community with notions of corporeality or physical space.

This matter of the dissolution of the material into the virtual has proved contentious, especially for those who regard the immediacy of the body as constituting an ethical imperative, 'the living genesis of cyberspace . . . the heart-beat behind the laboratory . . .' (Heim, 1993a: 80). Embodied interaction confounds the Platonic pretensions of mastery and purity in cyberspace. Heim refers (unattributed) to Emmanuel Levinas in an attempt to counter the Gnostic tendencies of virtual corporeality:

The living, nonrepresentable face is the primal source of responsibility, the direct, warm link between private bodies. Without directly meeting others physically, our ethics languishes . . . The face is the primal interface, more basic than any machine mediation. The physical eyes are the windows that establish the neighbourhood of trust. (Heim, 1993a: 75–6)

By equating reliable knowledge, and by imputation, moral integrity, with embodiment, Heim delineates embodiment as the distinguishing characteristic by which the dehumanizing antinominanism of virtual technologies can be resisted. Flesh marks the very province of our humanity.

Our virtual life in cyberspace paralyses our bodies. Cyberspacetime promises us liberation from the constraints of space, time and materiality. However, without the experiences of our bodies, our thoughts, our ideas, our ethics and politics must all suffer. We know ourselves and our world mainly because we live and move in the world through our bodies. (Nguyen and Alexander, 1996: 117)

In assuming the self-evidence and innocence of the corporeal, however, there is a suggestion of a foundationalism grounded in an unproblematic 'nature' which never changes, and is independent of representation. Such discussions depend on an appeal to an essentialist body as ground of ethics. Any appeal to the body as uncomplicated locus of identity is problematic when advanced medical and digital technologies displace the givenness of corporeality. In a post/human world, nothing is 'natural' and 'natural facts' cannot constitute, on their own, a reliable post/human ethic. So while many commentators regard the liberating potential of disembodied cyberspace as Cartesian, they collapse too readily into a kind of dualism that associates embodiment with an unproblematic naturalism that evades the very constructedness and pluriformity of embodiment. Just as Jacques Derrida castigated traditional philosophy for assuming that the spoken word is in some way primary and superior to writing (Bennington, 1998: 551–2), so these commentators fall into the trap of imputing a fixity and integrity to bodily (pre-virtual) experience. There is a tendency here, still, to bifurcate identity into 'real' self (bodily) and 'cyber' self (virtual) in an inversion of Platonic or Cartesian dualism, and to fall back upon a rather romantic vision of the unmediated encounter between humans who are assumed in no way to be constituted by technologies of any kind. We cannot allow such a rhetorical appeal to 'The Body' to bear the weight of ethical and political discourse without examining the ways in which 'our bodies' are themselves always already culturally, politically constituted.

Arguably, virtual reality still depends in a residual sense upon bodily proprioception and conventions of space, movement and perspective transferred from embodied experience (Vasseleu, 1994). Far from abandoning the body, forms of virtual interaction retain many of the conventions of face-to-face community. A form of synthesis occurs in which embodied experience transforms the technologies which mediate it, and vice versa (Argyle and Shields, 1996). It is notable that correspondence in cyberspace – such as e-mail or discussion groups – has evolved special symbols, or 'emoticons' to denote states of mind or indicate bodily expressions. Without necessarily taking place face to face, formal signs or representations of physical presence are still regarded as integral to proper communication. Indeed, their significance may be weightier given the need to underwrite virtual imprecision with further symbolic genuineness. Effectively, we might regard such conventions as an example of evolving 'netiquette' which seeks to elaborate a new symbolic of presence to compensate for the lack of actual corporeality as interactants leave physical 'traces' of themselves online. It is perhaps more appropriate, therefore, to think of cyberspace as a transitional state where the subject is both materially and digitally embodied (Green, 1997).

Was it a myth that we are separate from the machine, the computer; that it is cold, hard, circuitry that cannot be penetrated by the user? . . . Or could there be a way that individuals leave bits of themselves on the system, to communicate, and enhance our presence, leaving lures to catch the attention of one another even while we carry on with our day-to-day lives. (Argyle and Shields, 1996: 62)

Digital technologies may thereby be understood as retaining the vestiges of embodied presence, guaranteeing a measure of substance and referentiality, while also transforming and reconfiguring embodiment into many virtual corporealities.   189

The literal body may not be communicated within the Net, but it is possible to conceive of a 'multiplicity' of ways of being a *virtual* body that reflect a subjectivity which inhabits many levels of corporeal presence.

THE DISSOLUTION OF 'THE HUMAN'?

In the early modern era, anxiety was expressed at the realistic nature of automata and other artificial beings, for fear that audiences would be tricked into mistaking artifice for reality. Participation in cyberspace, whether it be in multiple user domains or virtual reality, evokes similar anxieties, that face-to-face (FTF) reality may prove less attractive and fulfilling than the simulated authenticity of the hyper-world. Does immersion in electronic media spell the end of FTF interaction, and will virtual technologies create a generation of atomized introverts and spell the disintegration of civic and collective cultures? This is a commonly expressed response to the explosion of digital technologies, and preoccupies much research and comment on the effects of cyberspace on individual psyche, on personal relationships and family and community life. Vivian Sobchack characterizes life on the Net as one of 'interactive autism' (1993: 574) where all contact with others is safe and sanitized. Cyberculture's dangerous allure rests in its potential to become a refuge from the complexities of the 'real' world:

Identifying with computers can be appealing on several levels in our fragmented postmodern existence. Vulnerable late twentieth-century bodies and minds turn to electronic technology to protect themselves from confusion and pain. Fusion with computers can provide an illusory sense of personal wholeness; the fused cyborg condition erases the difference between self and other. Additionally, a wholesale embrace of computerized existence can create a sense that one's messy emotions have been replaced by pure logic and rationality. (Springer, 1996: 128–9)

Alternatively, however, proponents of cyberspace argue that virtual integration harnesses participants' creative imagination, enabling them to create new personae and forge new kinds of communities, albeit mediated by electronic means of production. Sherry Turkle's research, some of it as participant observer, provides data for analysis of users' own understandings of their activities (Turkle, 1984; 1995). Many report that the opportunity to experiment with diverse identities in MUDs, virtual games or chat networks has enriched their understandings of themselves and others. A commonly reported effect is of 'cross-gender' identity, in which participants in on line discussion groups and in virtual sexual encounters adopt alternative gender identities or sexual orientation. The physical necessity of 'passing' as the member of another gender may preclude this kind of pretence in 'real' life; but the nature of virtual interaction opens new opportunities. Far from attenuating human intelligence, initiative and potential, this is a view of cyberculture as extending human creativity and sociability.

Turkle's research may reflect a bias towards those who express technophilic attitudes, but other researches bear out her broad conclusions (Roberts and

Parks, 1999). Just as social and geographical mobility in earlier phases of modernity enabled individuals to reinvent themselves, so cyberspace affords a much greater freedom to create new selves. While the self may be 'decentred' and multiple, participants tend to resolve this as being an enrichment rather than a dissipation of identity. If information on a Web page is often 'hypertextual', in which text or graphics are linked, or like Windows in which one can work in a software program by switching between parallel pages, then coherent identity is not unitary but multilayered. Sadie Plant argues that the World-Wide Web is spontaneously evolving as a decentralized, 'rhizomatic' matrix, with similarities to the complex interactions of the human brain. This, for her, is the future of artificial intelligence, tending towards a diversification of cognitive patterns, and of information as matrix rather than hierarchy or linear structure: 'Intelligence is no longer monopolized, imposed or given by some external, transcendent, and implicitly superior source which hands down what it knows – or rather what it is willing to share – but instead evolves as an emergent process, engineering itself from the bottom up' (Plant, 1996b: 204).

According to this view, identity in a cyberworld is fluid and negotiable: 'In this game the self is constructed and the rules of social interaction are built, not received' (Turkle, 1995: 10), closest to a conception of self as performative, where the electronic domain is the stage and where relationships and selves can be 'enhanced, augmented, changed, or erased' (Springer, 1996: 132). It induces a sense of a more fluid identity: one effectively 'performs' a *persona* which may be very different from one's own embodied 'real' self. Virtual interactions may simply accentuate the extent to which subjectivity has never been a constant. Just as various cybernetic and genetic technologies are dissolving the notion that bodies end at the skin, suggestive of physiologically (and psychologically) *porous* subjects, therefore, negotiations with the multiple domains of virtuality reveal *decentered* subjects for whom unitary identity is a cumbersome and irrelevant fiction.

Jean Baudrillard, reacting to the advent of new technological media, continues the theme of scepticism towards the 'real' essence enduring outer representation – linguistic or material – itself. His is a world in which, like cyberpunk's fictional world, truth and reality are indistinguishable from illusion or simulation. Classical Marxian materialist 'use-value' has been superseded by the ubiquity of commodification and 'signifying-value'. Digital technologies are capable of producing artefacts that are more 'real' than the authentic versions; the shopping mall, the amusement park, recreate a vision of pavement cafés and town squares that subsequently becomes the index for authenticity (Baudrillard, 1988: 171–2). This is an issue anticipated a generation before Baudrillard by the German-Jewish Marxist critic Walter Benjamin. Within a broadly Marxist analysis, Benjamin introduces the issue of how developments in technologies might affect social relations (Benjamin, 1992). Benjamin was ahead of his time in conceiving of technologies as media of organization and signification, as constitutive of social relations (and thus of alienated or authentic personhood) rather than as a mere tool. For Benjamin, innovations in the material basis of producing a work of art may well engender a transformation in the nature of the product itself. To abstract or distance the artefact from the material circumstances (and social relations) of its creation is to fetishize it, or reify the traces of its own fabrication. But in an age of mechanical or mass production, the old (and value-conferring) criteria of presence, originality, creative genius ('aura') are eroded (Wolin, 1994: 188–9). Technologies thus give rise to new aesthetic standards and facilitate new ways of apprehension. Benjamin gives the example of slow-motion photography which transforms what it records, to the extent that technologies now create and construct the very reality they supposedly depict (Benjamin, 1992: 211–30).

From a Marxist theoretical background, Baudrillard is attuned to this turn, and much of his work may be read in this spirit (Kellner, 1989). Technologies of reproduction and representation are nothing new; the eighteenth-century fashion for trompes-l'oeil and automata began the process of the copy as spectacle in its own right, attaining a value and novelty in itself. Nineteenth-century mass production enabled reproduction even in the absence of an original. The communication and information revolution of the late twentieth-century displaces the balance between 'copy' and original still further, as the manufacture of digitalized, virtual renditions of the real themselves become adopted as the 'original article'. In a time of the persuasive artifice of the simulacrum, therefore, technologies manufacture the conditions in which any distinction between the 'real' and the 'simulation' is confounded. This is the realm of what Baudrillard termed 'hyper-reality' (Baudrillard, 1983; Sarup, 1993: 165). Virtual technologies also enable reality to be synthesized so that the copy is more real than the original – if indeed there is an original. The representation of the real becomes the real, a kind of digital Marxism, in that Baudrillard is transposing notions of 'ideology' – a partial representation becoming accepted or imposed as the absolute account – into a digital context. The text (in this case, the binary codes of virtual technologies) 'decomposes' (Bey, 1998: 3) the material texture of the real and collapses the two together.

Emerging digital and biogenetic technologies thus represent fertile empirical opportunities for poststructuralism to speculate about the death of the (humanist) subject. For many poststructuralist philosophers, the doctrine of the rational humanist subject is not only an illusion but a repressive fiction. Many disciplines have developed critical perspectives which dismantle 'the self-sustaining subject at the centre of post-Cartesian western metaphysics' (Hall, 1996: 1). Mark Poster suggests that the Internet may be a fully postmodern medium insofar as it has realized Jean-François Lyotard's vision of a proliferation of local narratives to displace the 'grand narrative' of modernity. Experiences of working with advanced technologies of immersion and incorporation destabilize former notions of subjectivity as a rational unitary self in a bounded body. Digital communication is the vehicle not simply for the transport of a pristine self but is the very medium of its constitution (Stone, 1993: 85). This view, of the subject not pre-existing its environment, but constituted within the multifaceted technologies of its creation, is of course at the heart of poststructuralist theory. Poststructuralism advances a view of subjectivity as constituted within networks of language and power, resistance to essentialism and historicism, and places an emphasis on the *performative* nature of subjectivity – fluidity of consciousness, identity, embodiment and gender. 'The rational, autonomous individual who pre-exists society, as Descartes and Locke maintained, emerges after the critique by post-structuralists as a western cultural figure associated with specific groups and practices, not as the unquestioned embodiment of some universal' (Poster, 1996: 198).

The challenges of AI, from the Turing Test to Rodney Brooks's emergent, proactive intelligences, may suggest, similarly, a turn to performativity in understandings of the subject. There is no underlying 'essence', simply a series of representations, enactments or simulations. For Judith Butler, identity is invention and artifice (note how, already, there are connotations of machines and technologies) (J. Butler, 2000). There is no 'self' behind the expressions and performances of identity; rather, there is an insistent dissolution of all concepts of subjectivity and agency outside the enacted context of identity-as-verb.[7] We are only subjects insofar as our choices and actions are always already constituted within structures that embody certain patterns of power and difference, calling us into being within their discursive framework. Butler thus effectively reverses the modernists' equation of subject > agency > politics, into politics > agency > subject (J. Butler, 2000). This represents a departure in concepts of what it means to be human, offering a model of subjectivity as 'a radically contingent phenomenon *constructed* through social practices, rather than an expression of an inner nature *revealed* through engagement with the social' (McNay, 1992: 171, my emphasis).

Note, however, that this presupposes neither an essential human subject who will be effaced by technology nor one whose rational subjectivity will endure regardless of an embodied condition. It suggests instead a very different understanding of the self, in which subjectivity is under construction, in a state of becoming, within material, linguistic, scientific and environmental contexts.

Monsters in metropolis

The science-fiction genre known as cyberpunk, which first emerged in the 1980s, is in some respects a direct descendent of Lang's vision of the future in

---

[7] Equally, for Butler the established distinctions of second-wave feminism between 'sex' and 'gender' are a fiction. Rosi Braidotti's depiction of Butler's characterization of 'gender-as-a-verb' well captures this: gender is not a trait to possess (which we might term 'gender-as-a-thing') but is an aspect of an identity that is always already performative (Braidotti, 1994b: 67).

Metropolis. The urban wasteland characteristic of cyberpunk is similarly divided between the fabulously wealthy – liberated from physical imperfection, bodily confinement, even death, by advanced technologies – and the urban underclass. It is also distinguished by its dystopian mood, indicative of a loss of bearings satisfied by neither futuristic technophilia, technocracy nor nostalgic desire for the unadulterated body. The Canadian writer William Gibson is perhaps the best-known and most influential author in cyberpunk, especially through his trilogy *Neuromancer* (1984), *Count Zero* (1987) and *Mona Lisa Overdrive* (1990). Gibson is credited by some ('Frank', 1998) as coining the term 'cyberspace' to denote a digitally generated realm of 'consensual hallucination' (Gibson, 1984: 12), a dreamlike state of jacked-in consciousness. Gibson also refers to the 'Matrix', a global network of computer systems depicted as a sort of graphic database, reminiscent of early video games, especially in its assumption of data as graphically represented and in the almost ludic nature of bodily absorption into cyberspace.

Cyberpunk portrays varieties of post/human subjectivity as fully assimilated hybrids of technology and biology (Ross, 1991: 137–45). Human characters merge and interact with technologies, both in the form of information, especially the Matrix (the global network of all computer databases, in which information is represented graphically), and in the form of sentient beings (as in the various artificial intelligences and Dixie, a hacker who 'flatlined' and whose personality is digitally stored). The physical environment is either that of urban decay and social disintegration or computer-generated virtual worlds. Nature, unless heavily adulterated, does not feature in cyberpunk. In contrast to earlier science-fiction genres, Gibson's mood is dystopian and bleak. There is a suggestion in *Neuromancer* of post-nuclear apocalyptic, characterized by disparities between a rich elite with access to all the fruits of technology and a marginalized underclass, of hustlers, prostitutes, hackers and information pirates. Transnational corporations, the media and religious cults have replaced any recognizable body politic. There is precious little altruism, charity or political will. Personal relationships are transitory, exploitative or dysfunctional and secretive, panoptic corporate power dominates the public domain. Gibson's chief protagonists are anti-hero(ine)s, fighting the anonymity of the system from the margins, living rootless, precarious lives on the verge of subsistence. Digital technologies, especially computers, artificial intelligence and cyberspace are ubiquitous; and the technologization of society extends to its human members, in the form of prosthetic limbs and implants, chemical enhancements, cloned or genetically engineered persons, organs and body parts.

In a world where few bodies have not been enhanced in some fashion, there is a contempt for the dreary (and unnecessary) limitations of corporeality. The body is mere 'meat', to be abandoned in the 'bodiless exultation' (Gibson, 1984: 12) of the Matrix, or to be reconstructed, even cryogenically preserved, in order to evade physical decay and death. AI, especially in the sinister forms of the twin entities in *Neuromancer*, Wintermute and Rio, has developed to a sophisticated stage, but is subject to strict monitoring and limitation (by a force known as 'the Turing police'). It is possible, via digital means, to store the personality of those who are dead, and to experience vicariously the physical and emotional sensations of other individuals via a mechanism known as 'sim-stim'. Not only are the boundaries and the primacy of the physical body in question, therefore, but the integrity and uniqueness of personal experiences, thoughts and emotions.

Many of these ideas are echoed in later cyberpunk works. Pat Cadigan's *Synners* (1991) explores typical themes, such as the technological erosion of human distinctiveness, the confusion of reality and illusion and the totalitarian tendencies of corporate capitalism (Balsamo, 1997: 134). *Blade Runner* brings to life on celluloid a vividly Gibson-esque atmosphere, especially its depiction of a polyglot urban sprawl that reflects the syncretization of East and West into a new global consumerism.[8] The plot of *Strange Days* (directed by Kathryn Bigelow, 1995) centres on Lenny, a huckster who peddles virtual recordings of others' experiences for customers in search of vicarious thrills; and *Johnny Mnemonic* (directed by Robert Longo, 1995) adapted from a short story by Gibson (1997), features an eponymous hero who smuggles and pirates information by having it 'downloaded' into a prodigious and electronically enhanced memory. Non-human animals are not unaffected by the invasive metamorphosis of cyborg technologies, so *Johnny Mnemonic* also features an artificially enhanced dolphin, formerly employed by military intelligence but now ruinously addicted to the pharmaceutical cocktails used to enhance its abilities.

The sentiments of cyberpunk mount a rebellion against the increasing commercialization of science fiction, and in particular against the humanistic utopianism of much mainstream science fiction. Gibson's early short story, 'The Gernsback Continuum' (1986) is a clear protest against this genre, repudiating the technocratic futurism of Hugo Gernsback's progressivist popular science (see Chapters two and six). The assumptions of progressive/humanist science fiction are all upended, especially a faith in technological progress and human perfectibility.

This is a thoroughly dystopian vision, abandoning the progressivism of humanism and setting forth a prediction of the erosion and colonization of human freedom by inhuman technologies and impersonal corporate capitalism. In the bleaker cyberpunk worlds, where body parts, mercenaries, loyalty and human emotions are all commodities to be obtained at a price, it is only the imperatives of global capitalism via the cash nexus that determine the authenticity, or value, of anything. While some critics interpret cyberpunk's repudiation of the

---

[8] Note the portrayal of transnational corporations as reflections of the ascendancy of Pacific Rim capitalism (what before the recession of 1998 were known as 'tiger economies') and the attribution of Japanese influence on economic and popular culture. This is also a feature of *Body of Glass* (Yakamura-Stichen corporation counterposed to the utopian settlement of the freetown) and of the visual aesthetics of *Blade Runner*.

utopianism of earlier humanistic science fiction as an eschewal of any political values (Rosenthal, 1991) it may rather reflect 'a multitude of examples of the irrelevance of the integrity of the human' (Fitting, 1991: 302). Correspondingly, ethical responses derived from conventional forms of humanism fail to resonate. Certainly, this extinction of the human carries us further into a world of post/humanities in which attachment to humanism has all but vanished and in which space for new discourses of what it means to be post/human may emerge. In contrast to the humanist representations discussed in Chapter six, cyberpunk does not shrink from the prospect of technologies as constitutive media of 'human nature', nor does it insist that humanity is forged outside culture, untouched by technological or social context. Cyberpunk articulates an understanding of the post/human as always already mediated by material conditions, and in a world where humans and machines are indistinguishable, implicitly rejects any ethical stance founded on humanist conviction. If cyberpunk does not exhibit clear-cut fears or hopes at its possible post/human futures, it may be because there is no longer any way of telling where the regions of the monsters begin and end: 'What aspect of humanity makes us human? Our flesh? Our thoughts? Our handiwork? Where's that line over which lies inhumanity? The technology is us, man' (David Porush, quoted in McCarron, 1995: 264).

'Your Aura will not be your own'[9]

If poststructuralism provides a theoretical framework for the representation of the post/human beyond humanism, then the work of the Australian performance artist Stelarc may be one variant of the practices which attempt 'to activate alternatives and open up the range of human destinies' (Goodall, 1999: 168). Stelarc specializes in the spectacle of his own body interfacing with technologies. Among his various installations have been the attachment of a 'Third Hand' to his body (Stelarc, 1998: 119) and 'Internet Upload Event', in which electrodes stimulated by signals sent by a network of computers stimulate his muscles to move involuntarily. Stelarc's vision of the technologized future is one in which the defining motif is of the obsolescence of the human. 'THE BODY MUST BURST FROM ITS BIOLOGICAL, CULTURAL, AND PLANETARY CONTAINMENT' (Stelarc, 1998: 116). Stelarc equates this with 'the body shedding its skin' (Stelarc, 1998: 116), as various forms of prosthetic additions to the body actually transform the quality of embodiment, 'reconditioning' its circadian rhythms with cybernetic stimuli (1998: 117). Skin is no longer the boundary with the 'outside world' but the interface into the machinic. 'ONCE A CONTAINER, TECHNOLOGY NOW BECOMES A COMPONENT OF THE BODY' (Stelarc, 1998: 118), for as technologies transform embodied experience, so too what it means to be

[9] (Stelarc, 1999: 125).

human will be redefined: '[I]n the terrain of cyber complexity that we now inhabit the inadequacy and the obsolescence of the ego-agent driven biological body cannot be more apparent' (Stelarc, 1999: 122).

Although Stelarc has declared that he considers the body to be 'obsolete', he also argues that post/human development must be an embodied experience, not the 'meat-free' transcendence of transhumanism. If this is obsolescence, it is the dissolution of the naturalized body, and an exposure of the redundancy of an idealized body that, effectively, never existed.

A shifting, sliding awareness that is neither 'all-here' in this body nor 'all-there' in those bodies. This is not about a fragmented body but a multiplicity of bodies and parts of bodies prompting and remotely guiding each other. This is not about master-slave control mechanisms but feedback-loops of alternate awareness, agency and of split physiologies. (Stelarc, 1999: 120)

By exploring the porousness of the interface between humans and machines through his own bodily performances, Stelarc claims to enact a process of a technologically driven redesigned body (Atzori and Woolford, 1995). Yet although Stelarc speaks of a postbiological future, I also understand him to be insisting that whatever its composition, post/humanity will always be corporeal, and any future evolution must be proprioceptive.[10] In talking of a pluralism of embodied forms, Stelarc is hinting at a dissolution of humanism rather than a linear evolutionary trajectory.

Technology has always been coupled with the evolutionary development of the body. Technology is what defines being human. It's not an antagonistic alien sort of object, it's part of our human nature. It constructs our human nature . . . My attitude is that technology is, and always has been, an appendage of the body. (Stelarc, in Atzori and Woolford, 1995: 1)

Stelarc reveals embodied presence and auratic creativity as always emergent and technologically mediated. Stelarc's body, albeit merged with and assimilated into machines, is the focus of his art, which seeks not to transcend embodied materiality but to chart its limits and potentials. When Stelarc speaks of 'evolution', he means the potential of the digital and biotechnological age to reconceive post/humanities, envisaging radical incorporation rather than either the technological sublime of transhumanism or the striving for 'self-maintenance at any price' (Pearson, 1997: 232). Rather, the future is one of multiple possibilities. 'This is also the dilemma with a lot of science fiction, it postulates a sort of utopian ego-driven future as if this future already exists, whereas I tend to want to function in the way that, at any present moment there's a multiplicity of choices, and an infinity of possibilities' (Stelarc, in Farnell, 1999: 136).

---

[10] 'Proprioception is the body's so-called "internal sense" of its own position, both spatially, and in the relation of parts to each other' (Vasseleu, 1994: 160).

Like Foucault's genealogical critique, therefore, Stelarc may be seen as practising and embodying a journey of enquiry to the limits of human experience – be that surrender, postbiological corporeality or extremity – in order to interrogate the nature of human nature itself. Performativity precedes and iterates the meaning of embodiment and of human identity. The body is understood not as monadic entity or autochthonic organism but as a 'theatre of becoming' (Pearson, 1997: 230) in which Stelarc is dissolving the distinction between subject and object. In effect, therefore, such perspectives represent an era of 'posthumanism' in their radical questioning of the precepts of modernity concerning the self-evidence of the rational, autonomous, self-actualizing subject at the heart of the teleology of emancipation.

Stuart Hall suggests that amid the fluidity and decentring of the subject, the term 'identification' provides a useful alternative to terms such as 'identity' or 'the subject'. It suggests a pragmatic or process-orientated approach in the formation of selfhood. 'In contrast with the "naturalism" of this definition, the discursive approach sees identification as a construction, a process never completed – always "in process"' (Hall, 1996: 2). The very act of pushing concepts to the limits, of tracing their genealogies and usages, exposes their terms of reference. This does not necessarily negate the idea of the human subject, but is rather a call to place such concepts 'under erasure' (Hall, 1996: 1):

To think the subject and subjectivity differently is most certainly to give up on the unified, self-present and pre-given subject of modernity, but it need not suggest a permanent state of fragmentation. Contrary to certain construals of postmodernism . . . what I understand the deconstruction of the subject to entail is its constant re-formation in specific, local and temporary configurations. (Shildrick, 1997: 154)

Human subjectivity cannot be equated with a single privileged aspect, such as mental functioning. Rather, the mind and the self are themselves intertwined with physical and proprioceptive transactions. The subject is always an organic-technological body-in-relation, both creative agent and created subject within its changing environment. This is not intended to be an anti-humanist stance, but rather to argue that technologies must be seen as co-evolving and intermingled agents in the construction of the post/human. This is a shift from technophobia – and a view of 'technology' as monolithic, undifferentiated, beyond human agency – towards a reclamation of technologies as heterogeneous, and primary in their capabilities to mediate 'the human'. With that goes the view that there is no essential body. Poststructuralist philosophies have emphasized the extent to which the subject is not an ahistorical, enduring essence, but 'a crossroad of intensive forces . . . a surface of inscriptions of social codes' (Braidotti, 1996b: 3); and artists like Stelarc have demonstrated in their work the reconstitution of the human body-self within transformative networks of technologies of extension, incorporation and immersion.

The 'end of the human' need not necessarily entail a choice between 'impersonal deterministic technologized posthumanism' and 'organic unmediated autonomous "natural" subjectivity', but may involve modes of post/humanity in which tools and environments are vehicles of, rather than impediments to, the formation of embodied identity. It is conceivable that the effects of new cybermedia and their proliferating subcultures will rest not in the straightforward destruction, but in their 'complication of subjecthood, their denaturalizing the process of subject formation, their putting into question the interiority of the subject and its coherence' (Poster, 1996: 199, my emphasis). Such a bricolage of miscegenation, hybridity and assimilation defies and destabilizes models of the subject protective of an ontological hygiene of humanism.

CHAPTER NINE

Cyborg writing

Cyborg writing must not be about the Fall, the imagination of a once-upon-a-time wholeness before language, before writing, before Man. Cyborg writing is about the power to survive, not on the basis of original innocence, but on the basis of seizing the tools to mark the world that marked them as other. (Haraway, 1991a: 175)

The previous chapter closed with a discussion of the performance artist Stelarc whose installations demonstrated the creative possibilities of post/human subjectivity and embodiment. In the assimilation of digital and prosthetic technologies into his body, his immersion in loops of feedback and machinic agency and his relinquishment of humanist individuality, Stelarc is a quintessential *cyborg*. The cyborg, a hybrid of the cybernetic and the organic, straddles the boundary of human and technological and disrupts many taken-for-granted distinctions between 'natural' and artificial. In the feminist theory of Donna Haraway, the cyborg represents an exemplary figure of post/humanity, one for whom 'history has forbidden the strategic illusion of self-identity' (Haraway, 1992a: 329). I suggest in this chapter that we might regard Haraway's famous discussion of the cyborg, notably in her essay 'A Cyborg Manifesto' (Haraway, 1991a) as another form of *genealogical* reading of post/humanity, exposing the fissures and fault-lines present in appeals to fixed identities, 'transcendent systems' (Marsden, 1996: 6) and instrumental rationality.

Yet alongside that critical task, Haraway seeks also to articulate a new ethical and political sensibility grounded in the hybridity and contingency of the cyborg. Much of Haraway's work, from 'A Cyborg Manifesto' to more recent material, has deliberately repudiated representations of nature and humanity as unadulterated phenomena independent of technological 'contagion' (Penley and Ross, 1991b: 6). What I term 'cyborg writing' encapsulates this ironic and anti-utopian emphasis on Haraway's part, stressing the locatedness and contextuality of knowledge and refusing to adopt absolutized or purist responses to the ethical, ecological and political dilemmas of technoscience.

As we shall discover, therefore, Haraway eschews solutions and utopias that attempt to recall us 'to an imagined organic body to integrate our resistance' (1991a: 154), be that a rhetoric of unity with nature or totalizing politics founded upon the integrity of women's embodied experience. For Haraway the apotheosis of such strategies of romanticism and essentialism is the figure of the Goddess.[1] Yet while appeals to unreconstructed nature and women's experience are clearly inadequate for the formulation of a post/human ethic, Haraway's ready dismissal of notions of divinity associated with the Goddess echoes earlier one-dimensional evocations of 'transcendence', discussed in Chapter eight. While full of critical power, Haraway's cyborg writing is incapable of resolving the polarity between the technological/postbiological sublime (in the shape of the patriarchal sky-god) and a compensatory restoration of the 'chthonic' (Roberts, 1998: 282) integrity of the earth-goddess.

Cyborg manifestations

The term 'cyborg' was first used in a paper published in 1960 by two aeronautics experts, Manfred Clynes and Nathan Kline. Their paper speculated how techno-logical adaptations of physical functioning might enhance human performance in hostile environments such as outer space (Clynes and Kline, 1995). Subse-quently, cyborg technologies have been developed to restore impaired functions (such as prosthesis or implants), or modify existing capabilities, enhancing bodies into stronger, better, faster systems. Cyborg technologies are diverse. One summary identifies four types: the 'restorative', replacing lost or impaired functions; the 'normalizing' – restoring a creature to 'normality' (although the state of that normality is not interrogated); 'reconfiguring' – creating posthuman creatures or modifying existing faculties; and 'enhancing' or remaking bodies and consciousness into powerful or foolproof systems (Gray, Mentor and Figueroa-Sarriera, 1995).

Haraway's celebrated paper 'A Cyborg Manifesto' (first published in *Socialist Review* in 1985) articulates what she terms the 'blasphemy' (1991a: 149) of cyborgs as transgressive, hybrid creatures who destabilize the very categories on which Western scientific logic depends. Her vision encapsulates well the developments, already charted here, associated with a digital and biotechnological age, and how advanced genetic, digital, biochemical, cybernetic and mechanical techniques have made possible countless new configurations of the organic body through prosthetics, permanent or temporary modifications and therapies. Cyborgs inhabit a world simultaneously 'biological' and 'technological'. As living fusions of the

[1] I will refer to 'the Goddess' when discussing the figure evoked by thealogians and other goddess feminists. When referring to more general references to female deities, and to varieties of woman-centred spiritualities, I shall use terms like 'goddess feminism'.

human and non-human animal, the human and the mechanical and the organic and the fabricated they render transparent the 'leakiness' of modernity's boundaries between species and categories (Haraway, 1991a: 152–3).

Haraway considers cyborgs to be both literal and figurative beings, straddling fictional and scientific worlds, exemplifying the problematic distinction between the two. The cyborg tells of 'the inextricable weave of the organic, technical, textual, mythic, economic and political threads that make up the flesh of the world' (Haraway, 1995: xii). Thus the cyborg is a symbol, a metaphor – maybe even the prime representative – for post/human metamorphosis in a techno-scientific age. The cyborg also promises a renewal of relationship between humanity and what have been characterized as nature and technology, a greater intimacy and complicity with environment and artefact, in which human nature is no longer characterized through mastery and exclusion of its designated others. The cyborg, however, does not just symbolize the generation of new creatures. It serves as a metaphor for the deep dependence of Western culture on technologies, the increasing significance for late capitalism of the processing and circulation of information rather than production and consumption; and the malleability of human genetic constitution.

In order to understand the significance of the cyborg as rhetorical figure, it is important to acknowledge the influence of cybernetics on Haraway's thinking. As I argued in Chapter eight, cybernetics provides technoscience with a model of intelligence founded on the dynamics of command, control and communication, and of technologies as potentially self-regulating. This provides a system of representation that portrays human, organic and machinic life as sharing the same vitality. Haraway assimilates these principles of cybernetics into her preference for certain ways of representing the world. Identity is less a question of essences, but rather represented in terms of data flow, interfaces, coding and design, where the 'machines are surprisingly lively, and we ourselves frighteningly inert' (1991a: 152). The cyborg has no myth of origins, because it has no parents and, significantly, no divine creator.[2] It is self-creating and self-sustaining, a pastiche of components rather than an organic being with a beginning and an end. The cyborg is thereby released from both nostalgic yearning for lost innocence and from teleological justification: 'A cyborg body is no innocent; it was not born in a garden; it does not seek unitary identity and so generate antagonistic dualisms without end (or until the world ends)' (Haraway, 1991a: 180).

Just like cybernetic mechanisms, therefore, apes and cyborgs dissolve the distinction between the 'born' and the 'made' (Marsden, 1996: 10) and call into question the origins of the boundaries that underwrite Western modernity. This is, in a sense, a variant of genealogical thinking. Like Latour and Foucault, Haraway is exercising a hermeneutics of suspicion towards the axiomatic

[2] Haraway expresses antipathy to psychoanalysis for this reason. See Penley and Ross (1991b).

boundaries and categories by which modernity is ordered, and questioning the discursive and representational practices by which coherent accounts of 'human', 'animal' and 'machine' are engendered. Here Haraway displays a continuity between the 'Manifesto' and her work on primatology in *Primate Visions* (1992b) where she argues that the affinities and distinctions between species – be they apes and humans or humans and machines – are constituted by scientific representation, described in Foucauldian style as 'disciplined practices deeply enmeshed in narrative, politics, myth, economics, and technical possibilities' (Haraway, 1992b: 2). Thus simians, cyborgs and women all occupy the boundaries of modernity, positioned there to show forth the scientific and cultural narratives that 'determine what will count as knowledge' (Haraway, 1992b: 13).

In her attention to the processes by which categories are made – and core concepts such as 'sex', 'gender' and 'human' are reified – she reads against the grain of essentialism and naturalism. Her materialist analysis thus represents 'an argument for pleasure in the confusion of boundaries and for responsibility in their construction' (1991a: 151). Digital and biogenetic technologies are redolent with liberatory potential insofar as they confuse, deconstruct or render transparent the ontological hygiene underpinning Western modernity's classifications of power and difference, namely class, race, gender or species. Indeed, a Foucauldian analysis of the cyborg would argue that she can only ever be an ironic intervention in the genealogy of technoscientific power/knowledge. '"Cyborg" is a way to get at all the multiple layers of life and liveliness as well as deathliness within which we live each day' (Goodeve, 2000: 136). Haraway's theorizing is thus 'a patterned vision of how to move and what to fear in the topography of an impossible but all-too-real-present, in order to find an absent, but perhaps possible, other present' (Haraway, 1992a: 295).

So the myth of the cyborg concerns the transgression of ontological hygiene (or, more accurately, its exposure as fabrication), in which 'the boundaries of a fatally transgressive world, ruled by the Subject and the Object, give way to the borderlands, inhabited by human and unhuman collectives' (1992a: 328). Such territories, are, of course, traditionally the realm of monsters, and Haraway characterizes the creatures of the boundaries between species as 'monstrous' in their power to signify the limits and constructedness of the narratives of Western modernity (Haraway, 1991a: 180, 1992a).

From bestial monstrosities, to unlikely montages of body and machine parts, to electronic implants, imaginary representations of cyborgs take over when traditional bodies fail. In other words, when the current ontological model of human being does not fit a new paradigm, a hybrid model of existence is required to encompass a new, complex and contradictory lived experience. (González, 1995: 270)

For Haraway the cyborg embodies the ethical and political heart of critical studies of science, because it necessarily poses a question about the subjects,   203 objects and beneficiaries of scientific practice, and 'who gets to count as a rational actor, as well as an author of knowledge' (Haraway, 1997: 89). These epistemological questions are fundamental, not secondary, to scientific practice and accountability: 'My goal is to help put the boundary between the technical and the political back into permanent question as part of the obligation of building situated knowledges inside the materialized narrative fields of technoscience' (Haraway, 1997: 89).

The emphasis on the hybridity and contingency of life, derived from cybernetics, propels a wider moral imperative to realize a greater accountability for the effects and implications of technoscience: a responsibility born of awareness of its status as an integral aspect of our human creative intelligence, rather than an external demonized force beyond our control. Machines are essentially extensions of our embodied agency: 'The machine is not an it to be animated, worshipped and dominated. The machine is us, our processes, an aspect of our embodiment. We can be responsible for machines; they do not dominate or threaten us. We are responsible for boundaries; we are they' (Haraway, 1991a: 180).

Cyborg writing

How does Haraway's celebration of advanced technoscience differ from that of transhumanism, discussed in Chapter seven? Does she recognize the particularity of the privileges of the digital and biotechnological age, and the possibility that cyborg subjectivity is the province of a global minority? To celebrate the ubiquity and virtues of the cyborg without asking whether such an ambition merely enshrines the fantasies of the wealthy may be considered a version of the 'god-trick' Haraway demolishes elsewhere (Haraway, 1991b: 189).

Modest_Witness@Second_Millennium, Haraway's most recent book, addresses a series of related themes to do with representations of science on the threshold of the twenty-first century. The themes of hybridity and impurity recur as she argues that technoscientific representations depend for their power on miscegenation and boundary crossing, in the shape of anthropomorphized machines, cyborged animals and technologized nature[3] at the heart of contemporary global technoscience. In 'Cyborg Manifesto', too, she had already argued that the impact of the digital and biotechnological age – also an age of advanced consumer capitalism – entailed a global analysis, the chief ramifications being the feminization of poverty, the increasing power of multinational corporations, the rise of the homeworking economy and the commodification of formerly subsistence agriculture (1991a: 161–73). In this respect, Haraway has attempted

---

[3] For example, the image of planet Earth taken from an orbiting space station adopted as the symbol for an ecological campaign: 'Love your Mother'. Yet Haraway points out that without advanced technology, the very photograph that evokes feelings of solidarity with environmental protest could not have been taken (1992a).

to speak from her own specific vantage point amid the commodification and 'fetishism' of Western technoscience and simultaneously to trace their effects on the 'others' of the global economy, such as nature, women and colonialized peoples. (This means it is firmly a First-World, late-capitalist perspective, which is both a strength, in that it confronts the relative privilege of those who have access to the new technologies and a weakness, in that it does not easily afford a voice to those who do not.)

Haraway's very immersion in the power complex of such processes thus impels her to seek appropriate strategies of responsible and oppositional knowing, and which, in part, informs her conviction that such situated knowledge cannot start from anywhere but the inside – in this case, her own position as an educated, White, Western scientist. Haraway is aware of the abuses committed by modern technoscience and political expedience, yet she is not prepared to abandon completely the vision of enlightenment and progress she believes such institutions are still capable of embodying. Thus she speaks from a position that acknowledges her own privileged collusion with the institutions and ideals she wishes to oppose and rebuild: a relationship with the stories of late modernity in which changing the plot is a modest but substantial intervention.

The only thing my people cannot do in response to the meanings and practices that claim us body and soul is to remain neutral. We must cast our lot with some ways of life on this planet, and not with other ways. We cannot pretend we live on some other planet where the cyborg was never spat out of the womb-brain of its war-besotted parents in the middle of the last century of the Second Christian Millennium. (Haraway, 1997: 51)

Marsha Hewitt has queried whether Haraway's celebration of 'techno-mythology' (1993: 140) is desirable amid so much alienated anti-humanism of late capitalism. Hewitt argues for more of a commitment to humanism, not its dissolution, in a global context where subjugated peoples, and especially women, are still denied status as full persons. I suspect Haraway would argue that we cannot return to a rhetoric of 'full humanity' when advanced technologies have rendered such a term meaningless as fixed essence: it is only now a conglomeration of genetic manipulations, virtual realities and organic–technical couplings. A celebration of cyborg subjectivity – a hybridity that is only established in order to subvert understandings of identity as a fixed category – entails a refusal of metaphysics and nostalgia.

What I would term Haraway's 'cyborg writing', therefore, emerges from these concerns about political responsibility and complicity, and it has three dimensions. It is writing about hybridity and indeterminacy; but it is also writing from a position of immersion in advanced global technoscience. And thirdly, 'cyborg writing', despite its ironic overtones, represents an ethical position. Haraway writes partly in order to articulate renewed relationships between humanity, technology and non-human nature that do not rest on exploitation

205

and which acknowledge the complexities of Western technoscience. Relationships and representations for Haraway become 'articulations' speaking not of post/human distinctiveness and superiority over non-human nature, but of post/humanity's partnership and connectedness with other sentient beings (Haraway, 1992a: 324). For Haraway, myths of origin and teleological narratives abstract those who tell and receive them away from the material circumstances of their production and proffer tempting visions of nostalgic or utopian perfectibility. They function as an ideological opiate. By contrast, cyborgs exist 'outside salvation history' (1991a: 150), immune from Oedipal imperatives, nostalgic longings or teleological drives. Cyborgs are not part of any myths of unity and separation; rather they rework the categories of, and the relationships between, nature and culture. The non-innocence of all the categories – race, sexuality, gender, class, species – by which humanity labels itself, and describes its interests, enables a robust realism towards any hankering for an unsullied past or totalized future.

Haraway is unmoved by overarching themes of repression or radical rupture at the heart of personhood as portrayed in the various schools of twentieth-century psychoanalysis (Kunzru, 1997: 2–3; Haraway, 1991c). The ambitions of a cyborg are not those of a restoration of prelapsarian innocence, back to the garden, the helpmeet and the reconciliation with her Creator (with echoes of Frankenstein's creature). All these narratives in various ways represent the subject as a survivor of painful losses occasioned by the violence done to the psyche by virtue of the entrance into the symbolic or patriarchal order. But Haraway will have none of this. Her cyborg has endured no fall from primordial innocence, no Oedipal crisis, but also has no need, equally, therefore of a narrative of restoration. Cyborgs do not crave holism nor reunification with abandoned paradises nor maternal figures; instead, like Haraway's other favourites, the tricksters, cyborgs are restless nomads, 'neither female divinities nor biological mothers (neither Goddess nor Woman)' (Crewe, 1997: 900). So Haraway's narrative of redemption is about transition and change without loss, permanent wandering and transmutation without origins or ultimate destination. Refusing narratives of progress or millenarian apocalypticism, Haraway is more comfortable with

the absence of beginnings, enlightenments, and endings: the world has always been in the middle of things, in unruly and practical conversation, full of action and structured by a startling array of actants and of networking and unequal collectivities . . . The shape of my amodern history will have a different geometry, not of progress, but of permanent and multi-patterned interaction through which lives and worlds get built, human and unhuman. (Haraway, 1992a: 304)

Haraway deploys characteristically postmodern irony and anti-foundationalism while retaining a modernist's sensibility that fragments of emancipatory reason can still be retrieved. She argues that cyborgs are precisely about the delights taken in the affinities between human and non-human nature, involving a search for non-exploitative relationships and a celebration of 'permanently partial identities and contradictory standpoints' (1991a: 154). Haraway also uses the cyborg as a heuristic and fantastic device by which to interrogate the implicit gender biases inherent in the categories by which Western modernity orders itself. Like other feminist theorists of science, Haraway has exposed the degree to which the binary categorizations of modernity are implicitly gendered, so that the 'nature' against which science defines its task is linked to women, bodies and emotion. The categorization of 'woman' as non-human other in relation to the normatively human male is revealed as a construction of the discourse of the Scientific Revolution, responsible for the historical exclusion of women from scientific and public pursuits.

In terms of gender identity, therefore, Haraway speaks of the cyborg of her imaginings as female, 'a girl who's trying not to become Woman' (Penley and Ross, 1991b: 20). Her insistence upon the cyborg as gender-specific in this way enables her to envisage the potentially subversive effect of a female body (with its associations with organic matter, finitude, birth/death, the immanent, the non-rational) fused with electronic, prosthetic and cybernetic parts, of a real 'monster' transgressing boundaries and therefore challenging the symbolic dichotomy between things of nature (bodies, women) and things of culture (machines, minds, masculinity). By implication, Haraway's advocacy of the cyborg also renders the possibility of women's interaction with technologies more feasible by exposing the fiction of ontological gender difference grounded in metaphysical affinities between women and nature. The desired effect is to embolden women – formerly enclosed in discourses of non-rational and non-technical – to think of themselves as entitled to enter the world of technoscience and to counter a sexual division of labour that has often denied women access to rationality, the public sphere and technologies. The romanticized goddess of an unreachable 'nature' proves equally inadequate to address the digital and bio-technological age in which cyborg imaginings are already a reality.

One of the most important implications of 'A Cyborg Manifesto' is, therefore, its opposition to women's historic exclusion from science and technology. Cyborgs straddle the boundaries of nature and culture, revealing them to be a construction; they embody fusions of organic, mechanical and electronic life, thus challenging the association between technoscientific pursuits and gender difference. By placing all of Western culture – and especially women – in a position of engagement and complicity with technologies, Haraway resists the gendered division of labour that has forbidden women a stake in rationality, scientific innovation and public discourse. However, Haraway's characterization of the cyborg as transgressive of traditional gender stereotypes is by no means the only depiction. Her invocation of this quintessential post/human ironic figure cannot remain uncontaminated by other representations circulating in popular culture which serve to reinforce rather than subvert the excesses of mastery and domination.

Comics, films and novels (such as the *Terminator* and *Robocop* films, the *Borg* species featured in several of the later *Star Trek* television series and the genre of 'cyberpunk' novels most famously represented by William Gibson's *Neuromancer*, feature many kinds of hybrid human-machines (Gray, 1995a). Many of the most popular representations of cyborgs appear as hypermasculine killing machines, more *Marvel Comics* than cyberfeminism. Their appearance as armoured prostheses, with physical and mental powers exponentially enhanced by various technologies, betray a transhumanist craving towards invincible, omnipotent post/humanity. For the hypermasculine Terminator and Robocop, and many of Gibson's protagonists, prosthetic implants and enhancements, or the disembodied 'high' of cyberspace, represent technologized means of escaping the vulnerability of embodiment, as discussed in Chapter seven. Similarly, the 'recovering' Borg in *Star Trek: Voyager*, Seven of Nine, is ambivalent about abandoning the supposed perfection and omniscience of the cyborg collective for the bewildering contingency of human individualism (see Chapter six). A perennial feature of the representation of these varieties of cyborg, therefore, is that of the tension between their human and technological qualities, exemplified in the struggle between body and machine, or emotion and rationality.

One depiction in which technologies promise invincibility is *The Terminator* (directed by James Cameron, 1984) featuring a cyborg, played by Arnold Schwarzenegger, who is a ruthless hired assassin. The fact that the Terminator has been sent back in time by a race of artificial intelligence bent on annihilating the last vestiges of human resistance augments the hint of technologies as threat to human integrity. Schwarzenegger's own background in professional body-building – a sport which has for decades striven for its own kind of cyborg bodies with its celebration of bodies modified and enhanced to extremes by weight-training, nutrition and, it is widely rumoured, widespread abuse of anabolic steroids – reinforces his casting as the superhuman creature (Balsamo, 1996; Goldberg, 1995). The Terminator is pure body, and yet artificially enhanced in a way that puts him on a different plane to ordinary organic humanity. His chosen uniform, that of leather biker, also intends to signify tough macho masculinity, but it also carries with it a trace of the leather bars of gay subculture. In adopting such an exaggerated pose of hypermasculinity, Schwarzenegger demonstrates his own divorce from 'naturalism' and hints at a masculinity that betrays its own artifice.

In the sequel, *Terminator 2: Judgement Day* (1991), the identity of the Schwarzenegger character is more ambivalent. In a reversal of the previous plot, the Terminator is now fighting for the opposite cause, in defence of humanity against the machines. His mission in the first movie thwarted and destroyed, Schwarzenegger has been reprogrammed and sent back in time by the leader of the human resistance to protect an earlier version of himself as a young child from another assassination attempt at the hands of the race of artificial intelligences. In *Terminator* 2 there is a greater fluidity of gender roles. Mary Connor, the mother of the boy radicalized by her earlier brush with murderous conspiracies, has acquired a more muscular physique, a familiarity with an arsenal of weapons and associations with freedom fighters resisting US neocolonialism. Her 'harder' body and tougher attitude contrasts with Schwarzenegger's Terminator, more fully characterized in the sequel, who – in keeping with the greatest monster/alien convention ever – wishes to know what it means to be fully human. The growing bond between Mary's son, John, and the cyborg, and particularly the Terminator's gentle protectiveness, leads Mary to reflect ironically that the cyborg may be the best father figure John could hope for (Fuchs, 1995: 291). The Terminator's final act, of self-sacrifice in protecting John, is both reminiscent of Asimov's Three Laws of Robotics (see Chapter five) – of a machine programmed to protect at all costs – and is yet the point at which the cyborg can most genuinely express his emergent humanity. The final scene, as Schwarzenegger's body sinks to its demise in a molten vat, shows the very human gesture of a hand raised in farewell, an image far removed from that of the technologized killing machine.

While glorifying advanced technologies and featuring scenes of extreme violence in ways similar to the *Terminator* movies, *Robocop* (1987) and its sequel *Robocop* 2 (1990) contain greater ambivalence, even irony, at the prospect of cybermasculinity (Fuchs, 1995). After Murphy, the central character, is wounded in a sadistic attack, he is transformed into a cyborg. The technologies conform to expectations of a steely, muscular, militaristic robot, heavily equipped with firepower. As Robocop, the technological enhancements provide him with the wherewithal to avenge his own humiliating near-fatal assault. Yet there is a deeper level of critique. The viewer is inducted into a putative cyborg subjectivity from the very beginning of the film, as Robocop's consciousness – and his evident treatment as object, freak and commodity – begins to emerge. Despite his grotesque bodily invulnerability, Robocop's superficial toughness conceals a deeper fragility. He is haunted by the trauma of his attack and transformation. Once more, therefore, technologies mark his superiority and strength, but this is tinged with a sense of the irrevocable loss of his humanity (symbolized by shadowy memories of his home and family). Although cybernetic and prosthetic enhancements have saved Murphy's life, they are ambivalent blessings. The city in which Murphy/Robocop serves is an urban wasteland, rife with crime and violence; and, in a further reflection of classic cyberpunk themes, corporate capitalism is portrayed as sinister, secretive and voracious. The Byzantine 'Omnicorp' corporation, which owns and controls Robocop, is eventually exposed as the real villain of the piece. The masculine cyborg, while ostensibly adopting the axioms of technophilia, also articulates anxieties about the relationship between human autonomy and technology. There is an ambivalence, for while technologies can protect the vulnerable body, and help those who wish to do so to enhance strength or transcend finitude, the threat of technologies out of control – an alien, dehumanizing force – is also present.

Such representations may, in their ways, articulate their own share of ambivalence and irony, their inflated displays of invincibility overcompensating for the fact that such physical strength and muscularity is redundant in a world of microcircuitry in which small is likely to mean quicker and more powerful (Springer, 1996: 111). They may maintain the appearance of impenetrability and masculine resistance to technologies' immersive and incorporative faculties; but their strength is anachronistic. Beneath the masculinist technophilia there may lurk a deeper anxiety about the potential of technologies to devour, invade or undermine the sovereign self.

Haraway arguably pays insufficient attention to these alternative readings of the cyborg and to their persistence – and sophistication – as cultural representations of the post/human. While she acknowledges the cyborg's past life in the military-industrial complex – because the theme of complicity with technoscience informs her work throughout – she might have recognized that popular culture perpetuates transhumanist values, however open to parody and subversion they may be. Therefore, Haraway cannot claim a monopoly on cyborgs, or assume that they are innocent of contrary readings. They can only be used as a heuristic device to think beyond the polarities of technophilia and technophobia in the service of genealogical critique and fabulation.

Since the 'Manifesto' Haraway has developed her theme of responsible knowledge and representation as facilitating ethical scientific practice. Her concept of 'situated knowledges', first discussed in *Feminist Review* in 1988, criticizes classical epistemologies that appeal to objectivity and universalism as akin to a 'God-trick' that cloaks its own particularity beneath a pretence to universal truth-claims, a stance that is not only impossible, but a dangerous deception (Haraway, 1991b; 1994). Haraway calls for an embodied situated knowledge – a theme that owes much to feminist standpoint epistemology – which grounds its legitimacy claims in the partiality of vision that makes no pretence to its own non-innocence.

The grand narrative of salvation history is something Haraway has always sought to subvert in the figure of the cyborg, although she affirms once again the value of modelling political solutions on the 'modest interventions' (Vines, 1997) of a capacity to represent the world in a non-abusive fashion. This refusal to abstract herself from the complexities of technoscientific culture, coupled with a conviction that there is no such thing as a neutral vantage point from which we can engage with the world, requires Haraway to eschew any notion of a transcendent or omniscient salvation figure or narrative.

Cyborgs or goddesses?

It's not just that 'god' is dead; so is the 'goddess'. (Haraway, 1991a: 162)

Haraway makes a final statement at the close of 'A Cyborg Manifesto' which has been much-quoted. It is worth citing in full:

Cyborg imagery can suggest a way out of the dualisms in which we have explained our bodies and our tools to ourselves. This is a dream not of a common language, but of a powerful infidel heteroglossia. It is an imagination of a feminist speaking in tongues to strike fear into the circuits of the super-savers of the new right. It means both building and destroying machines, identities, categories, relationships, space[s,] stories. Though both are bound in the spiral dance, I would rather be a cyborg than a goddess. (Haraway, 1991a: 181, my emphasis)

It is significant that the title of Haraway's piece echoes Marx and Engels' *Communist Manifesto*, published in 1848. Haraway writes from a socialist-feminist standpoint, and shares the secularism of much post-Enlightenment thought. However, her Roman Catholic upbringing constantly resurfaces in her frequent albeit ironic uses of religious imagery. Look again at the references to charismatic religion, the Moral Majority – even to Starhawk's *Spiral Dance*. Within a Marxist or post-Enlightenment framework, the goddess – like any religious or metaphysical system – is seen as an impediment to the flourishing of human autonomy. In interviews Haraway has argued for strong affinities between Christian theology and techno-scientific work in their tendencies to proclaim apocalyptic and totalizing narratives (Penley and Ross, 1991b; Goodeve, 2000). For her, religious imagery represents one of the totalizing narratives of ultimate resolution and closure which is defied by the brave new world of which the cyborg is the metaphor and the herald. In an echo of the technological sublime of transhumanism and technochantment, religion once more legitimates a flight from finitude, decay and particularity.

Haraway's evocation of the Goddess as 'other' to the cyborg represents the former as a figure who tempts us to invest our energies in other-worldly visions, exemplifying a tendency to see power, creative will and most of all moral agency not in human efforts but in some heavenly, abstract realm. In this respect, in common with the overwhelming mood of second-wave Western feminism, Haraway is the heir/ess to Enlightenment secularism, in which religion is regarded as an oppressive and diversionary tool of patriarchy:

If women allow themselves to be consoled . . . by the invocation of hypothetical great goddesses, they are simply flattering themselves into submission . . . Mother goddesses are just as silly a notion as father gods. If a revival of these cults gives women emotional satisfaction, it does so at the price of obscuring the real conditions of life. This is why they were invented in the first place. (Carter, 1982: 559)

Against the cyborg's virtues of immanence, bodily and technoscientific hybrid-ity, the Goddess's putative transcendence, immateriality and disengagement stand

211

for everything negative and retrogressive. Contemporary champions of the God-dess herself would sympathize with Haraway's antipathy to such a deity on the grounds of her putative abstraction and other-worldliness, if transcendence is taken to denote a model of God who surpasses the contingencies of material being.[4] In an essay which celebrates the end of dualisms, however, Haraway's dichotomy of cyborg/goddess seems strangely incongruous. It suggests one final – and overlooked – ontological boundary, that between heaven and earth. But what if, as well as questioning the distinctions between humans/machines, human/non-human animals and humans/nature, there is a way of deconstructing the boundary between the spiritual and the material, immanence and transcend-ence? It is important to consider whether Haraway has unconsciously reified these categories and failed to see how they too might be contingent upon the strategies of translation and purification that characterize the rest of modernity.

Haraway's association of goddesses with world-denying ideology would be roundly rejected by many goddess feminists who argue that, far from reinforcing an (inverted) transcendent deity whose heavenly authority serves to sublimate political change, the Goddess overturns a system in which 'the best things about the world are somehow not a part of the world' (Goldenberg, 1995: 158). Having emerged in parallel with 1970s second-wave Christian feminism, goddess feminism – sometimes known as 'thealogy' – is now a thriving inter-national movement with many different emphases (Raphael, 1996b; Lunn, 1993; McCrickard, 1991). Many goddess feminists trace their devotions back to prehistoric civilizations in which the veneration of resplendent female figures mirrored pacific, egalitarian societies and where women were partners and not subordinates to men (Gimbutas, 1989). Nelle Morton rejects both the entity 'out there' or the prehistorical figure 'back there' as adequate expressions of Goddess (Morton, 1989: 111). For her, the Goddess is a life-giving force, one who restores to Morton her selfhood and who has 'ushered in a reality that respects the sacredness of my existence' (1989: 115). But this Goddess is within and among women's relationships of healing and affirmation, and is a metaphor for such an 'inner power and integrity' (1989: 117). This representation of the Goddess is more of a metaphor for women's energy and creative power, suppressed for so long by patriarchy and misogyny (Christ, 1979). However conceived, the Goddess of these contemporary thealogians is a very different deity from the patriarchal 'sky-god' of traditional Christianity. The Goddess is conceived as an immanent, intimate presence: '[R]ather than putting us in touch with a "changeless" God who stands above the world, Goddess rituals connect us to a divinity who is known within nature and who personifies

---

[4] Indeed, many other feminist, process and womanist theologians who would not identify themselves as goddess or post-Christian feminists would also view this model as a distortion and, like goddess feminism, would emphasize the immanence of God as one whose creative being is intimately tied up with the material world.

change . . . Not focused on life after death, Goddess religion calls us to hallow the cycles of birth, death and regeneration in this life' (Christ, 1997: 30).

Thealogy also stresses women's spiritual empowerment through ritual practice, such as the veneration of images and the creation of sacred space. These rites are understood to invoke gynophilic forces to facilitate change (Raphael, 1996b). A preference for *orthopraxis* (ritual practice) over *orthodoxy* (doctrinal teaching) further renders goddess spirituality necessarily concrete, immediate, embodied and material. It is less about worshipping an object limited to a particular place and time, than affirming the immanence of a sacred power whose energy animates the entire cosmos: 'To include women means to recognize the physical contingency of all thought and all creation . . . When theology becomes thealogy, the metaphysical comes home to the physical' (Christ, 1997: 160).

The Goddess thus symbolizes a conviction that humanity is part of a web or matrix of life. Images of the cosmos as the body of the Goddess conceive of the entirety of creation as sacred. There is no dualism of sacred and profane, because the whole ecosystem is connected to the divine:

The symbols and rituals of Goddess religion celebrate our connection to the cycles of the moon and the seasons of the sun and our participation in the mysteries of birth, death, and renewal . . . Goddess images resacralize the female body, enabling women to take pride in our female selves, encouraging men to treat women and children with respect and to acknowledge their own connection to the life force. (Christ, 1997: 165)

Far from being a denial of the material world, goddess spirituality locates divinity as encountered and affirmed in (female) bodies and senses which are the ways in which human beings encounter the world. It is unlikely that goddess feminists would appreciate the characterization of the Goddess as other-worldly abstraction, perceiving her as a being whose body is the earth, and whose presence imbues and sanctifies material life.

[T]he Goddess is the power of intelligent embodied love that is the ground of all being. The earth is the body of the Goddess. All beings are interdependent in the web of life . . . The symbols and rituals of Goddess religion bring these values to consciousness and help us build communities in which we can create a more just, peaceful, and harmonious world. (Christ, 1997: xv)

THEALOGY AND TECHNOPHOBIA

Thealogians may be able to defend goddess feminism against Haraway's indictment of other-worldly spirituality. Goddess spirituality locates the problem not in religion *per se*, but in patriarchy. If the 'maleness' of God is assumed to indicate a transcendent, dispassionate, disembodied divinity, then 'Goddess' symbolizes connectedness, solidarity and immanence. Goddess spirituality is clearly a wellspring from which political engagement of a sort is drawn, in the form of ecological sensibilities. By associating goddess spirituality with embodiment, the earth and nature, thealogians embrace a materialist reading of divinity and

213

transcendence; but in their appeal to an unreconstructed 'nature' and 'women's experience' they fail to address the very constructedness of these categories.

Whereas the cyborg's explicit transgression of traditional gender boundaries challenges such identifications of women with nature, embodiment and affectivity, goddess feminism's gender analysis is deficient by comparison. The Earth, women's bodies and the integrity of an imagined realm of unadulterated nature are regarded as the springs of authentic and redemptive experience, but at the expense of reinforcing gender dualisms that deny women access into the world of culture, knowledge and technology. In its inversion of the dualisms of nature/female/immanence and culture/male/transcendence, goddess feminism risks replicating rather than subverting gendered ontologies. Goddess feminism continues to locate women in the realm of a romanticized and nostalgic 'nature'; but this identification sanctions women's exclusion from the public domains of technoscience and cyberculture (Kunzru, 1997: 3; Penley and Ross, 1991b: 14). In conceiving of an alternative figure to the patriarchal 'sky-god', goddess feminists are indicted as resorting to an essentialized 'earth-mother'.

The association of the Goddess with an idealized, reified 'nature' may also be problematic. Christ's Goddess is a deity of beautiful nature, not compromised nor appropriated adulterations of nature/culture. Even within feminist theo/alogy, however, the association of feminine deities with maternal, pacific and nurturing qualities has disturbed many who regard this as a reinforcement of patriarchal gender differences. Womanist theologians in particular regard the benign mother god/ess as a sanitization of the ambiguities of real mothering (Raphael, 1996a). It means women and men in search of non-patriarchal images of human and divine parenting metaphors hear only about God the Mother as

nurturer, healer, caretaker, peacemaker, as though no other attributes were permitted God once she was fitted with a female pronoun . . . these definitions narrow God's role to the traditional feminine virtues: as though the Goddess were God's wife and our mother and her job was to clean up after all of us, and console us too. (Madsen, 1989: 103)

The tendency of women to project only good attributes on to the Goddess, and to evade questions of the ethical ambivalence at the heart of creation, is perhaps understandable given the betrayals and abnegations foisted on women by a patriarchal Father God. Nevertheless, the divine may also be a terrible force, beyond human morality; and the reality of war, disease, death and suffering is not something that can easily be effaced by the invocation of a nurturing Mother God/ess[5]:

[5] God/ess is a way of representing a gender non-specific deity (Ruether, 1992). At this stage in my argument, however, the gender of sky-gods and earth-mothers – stranded either side of a dichotomy of transcendence/immanence – is less urgent than the impending deconstruction of these binary categories themselves.

However certain one may be that one is loved by some presence in the universe . . . that same presence will kill us all in turn, will visit our lovers with sudden devastating illness, will freeze our crops, will age our friends, and will never for one moment stand between us and any person who wishes us harm . . . There is no escape . . . We may refute the notion of God as a punishing presence, always remote and forbidding. But, those obstacles gone, we come face to face with the essential dilemma, the vertigo, the horror of all ethical theism: we are more ethical than God. Given the power to make a world, we would never have made this one. (Madsen, 1989: 104)

Janet McCrickard detects harmful tendencies in simplistic religious narratives of the entry of evil into the world and in a dualistic (even Manichean) world-view in which reality is divided into a hierarchy of good and evil. For goddess spirituality, such a hierarchy juxtaposes masculine and feminine, assumed to be ontological categories within which, in a reversal of traditional patriarchal dualisms, the masculine pole is demonized while the feminine is benign and biophilic. Goddess feminists argue that thealogy begins in experience, and they delight in challenging the Enlightenment conception of knowledge as ruled by criteria of objectivity and rationality (Christ, 1997: 31–49). But McCrickard laments both the crude dichotomies of such analysis as well as its abrogation of the powers of reason, which is associated with patriarchal and masculine qualities.

While this valorization of experience and suspicion of reason is a valuable corrective, the danger comes when as a result women deny themselves a stake in rational thought. Critics of thealogy have pointed out its lack of rigour, as for example over the issue of valid historical evidence. Pam Lunn castigates 'the magpie-like eclecticism, and the often uncritical syncretism of much of the goddess-spirituality movement' (Lunn, 1993: 24). McCrickard concludes that the prospects for constructing a serious or critical scholarship around the study and practice of feminist spirituality therefore dissolve into 'pseudo-scholarship, a disastrous mish-mash of postulation, occult assertions, compounded errors, unfounded conclusions, inconsistency and wishful thinking . . . Women have deprived themselves of the means to critically evaluate their own ideas' (1991: 65).

McCrickard concludes, like Haraway, that such suspension of women's critical faculties is a disastrous sublimation, a diversion from the path to real feminist consciousness or lasting change. The nostalgic – but ultimately futile – longings of the goddess revival are merely one more opiate, serving to distract the powerless and oppressed who 'often resort to the occult as the only means of power and importance left to them . . . While fundamentalisms of every kind do indeed grant security and meaning, in doing so they close possibilities for women instead of opening them' (McCrickard, 1991: 65).

Romanticization of nature and flight from reason do nothing to bring us into the sort of robust engagement with techoscience that is at the heart of Haraway's vision. Her thesis is that science will not deliver the brave new world and no external force, technological or divine, will intervene to help us. She is not enamoured with the 'Big Science' of enterprises such as the Human Genome Project. What is celebrated is 'the pleasure of being at home in the world, rather than needing transcendence from it' (Penley and Ross, 1991b: 17); in other words, a cyborg whose immune system is allergic to goddesses, and a strategy in which humanity must learn 'to *redirect* technology to realize human and environmental values' (Barbour, 1993: 24, my emphasis) and not reify it. 'Cyborg imagery . . . means refusing an anti-science metaphysics, a demonology of technology, and so means embracing the skilful task of reconstructing the boundaries of daily life, in partial connection with others, in communication with all our parts' (Haraway, 1991a: 181).

Haraway thus strives for a more elaborate representation of 'nature' and inevitably finds herself at odds with much of the conventional rhetoric of ecofeminism (1991a: 174). 'Techno-realism' must replace 'phobic naturalism' (Penley and Ross, 1991b: 6). Similarly, her discussion of psychoanalytic and radical feminism in 'A Cyborg Manifesto' indicates her rejection of any totalizing accounts of women's oppression (1991a: 157–61), in common with many post-structuralist feminists' deconstruction of the unified, universal subject. Insofar as goddess spirituality makes a similar appeal to totalizing and prescriptive analyses of women's experience – in thealogy's case, of women's uncomplicated unity with pristine nature – Haraway's antipathy to goddess figures is unsurprising.

As we have seen, the cyborg's transgression of the boundaries exposes its artefactual and constructed character. Haraway favours models of nature/culture that acknowledge human agency in the making of the boundaries and interrelationships, and she seeks ways of acknowledging the integrity and complexity of nature's own agency as 'inappropriate/d Other' (Haraway, 1992a: 295). Arguably, though, she too evades questions of the destructive power and caprice of nature, and the necessity behind the human struggle to create habitable environments within hostile 'natural' conditions. Despite the *alterity* of in/appropriate/d nature, Haraway assumes its benevolence, like the rest of existence. There is thus no need for anything to counter the evils of exploitation and domination. In subsistence cultures, however, the constant presence of 'nature' as trickster is far from benign. Neither goddess feminism nor cyborg writing, arguably, engages adequately with the malevolent otherness of the forces we call 'nature'.

Nevertheless, Haraway's work represents a commitment to an epistemological and political project to develop scientific, ecological and social practices that represent 'nature' not as objectified Other but as 'Itself' (1992a: 298), an agent and partner in the relationship. But just as Haraway's evocation of nature as 'inappropriate/d Other' (1992a: 295), co-existent with humanity yet possessing its own agency beyond human appropriation or domestication, offers a reality beyond human mastery, it is possible that the divine too is more than the sum of human imaginings. We need to return to Haraway's unresolved dualism of transcendence and immanence, sacred and secular.

The 'crossed-out God'

For Haraway, the ubiquity of human/technological assimilation and the complexities of nature as both artefact and inappropriate/d Other mean that the problematic affinities between humans, machines and animals cannot be resolved by the 'cheap grace' of a blissful reunion with the maternal womb of nature. Rather, she is concerned to rethink human relationships with non-objectified others. While Haraway's scepticism finds a legitimate target in thealogy's gender politics, in other respects her opposition between human and divine, or (technologized) earth and (immaterial) heaven itself rests upon unexamined constructions of 'religion' and 'transcendence' that owe their origins to the Western Enlightenment. Arguably, though, this dualism informs all others. A metaphysical – or even ontotheological – understanding of human nature which perpetuates a final frontier between the material, embodied world and the spiritual, immaterial realm of supposed perfection and invulnerability retains damaging and exclusive associations of technoscience with transcendence, disembodiment and aggrandizing abuses of power.

Yet the frontier between humanity and divinity is, arguably, as much a product of modernity as the problematic objectification of nature, and the reification of gender difference – all categories that Haraway is all too ready to deconstruct – so why not the religious symbolic of modernity? Above all, the birth of modernity, and especially humanism, is premised on the creation of alterity in the guise of non-humans, God and 'nature', those things which have to be excluded from the coherent master category of secular humanism. In elevating human reason and human agency, modernity thus rests on what Bruno Latour calls the 'crossed-out God, relegated to the sidelines' (Latour, 1993: 13). The secular evacuation of religion in the name of purification, the separation of the natural and supernatural and the elevation to the heavens of a transcendent, immutable God are necessary parts of the logic of purification and translation. The invisible guarantor of predictable laws, discernible in the ordered patterns of creation, but distanced from intervening in the universe (as for example in Deism) poses no serious threat to scientific freedom. Those antimoderns fearful of the world losing its spirit do not realize the artifice with which disenchantment has had to be assembled. The categories of 'transcendence' and 'immanence' are, for Latour, as much the by-products of modernity's strategies of purification as anything else:

Spirituality was reinvented: the all-powerful God could descend into men's heart of hearts without intervening in any way in their external affairs. A wholly individual and wholly spiritual religion made it possible to criticize both the ascendancy of science and that of society, without needing to bring God into either. The moderns could now be both secular and pious at the same time . . . (Latour, 1993: 33)

Is it enough for cyborg writing merely to embrace contingency and complicity in a mood predominantly suffused with irony? Haraway's cyborg writing

217

adopts a postmodern veneer in its embrace of the hybridity and contingency of technoscientific culture but in other repects 'it remains all too modern' (Oliver, 1999: 340). An alternative approach might be to embrace the symbiosis of nature and culture while facilitating a critical imagination to reconfigure the categories of 'transcendence' and 'immanence'. Not transcendent in the sense of other-worldliness and immateriality, but as in critical, oppositional, visionary or utopian.

Once more, therefore, a discussion of the significance of the post/human has encountered the recurrence of themes associated with immanence and transcendence, divinity and spirituality. It is time to attempt some ways forward, and I want to begin with Haraway's own unresolved utopianism, myth and religious referents. Haraway's cyborg ethic celebrates the radical immanence and contingency of post/human becoming; but critics have observed that Haraway's eschewal of 'salvation narratives' deprives her of a moral or mythical imagination to transport her beyond the confines of humanism (Hewitt, 1993; Oliver, 1999). Ultimately, she fails to articulate a posthumanist sensibility that embraces a proper symbiosis of nature and culture yet points to or facilitates a wider vision that enables the cyborg to transcend its own irony. It is this aporia of critical imagination that renders her cyborg writing unsatisfactory. Either we fall back into the arms of humanism (the charge levelled against Haraway by Oliver and Hewitt) or we must look somewhere else.

Reading against the grain of Haraway's work, excavating the unacknowledged references without which her work would have no coherence, one is struck by the persistence of religious, utopian and moral imagery, in spite of her own disavowal of eschatological or teleological narratives.[6] More recently, however, Haraway has acknowledged the influence of the American Catholic Left upon her early political and theoretical ideas (Goodeve, 2000: 13; 141; Kunzru, 1997; Haraway 1997: 2–3): 'My deep formation in Catholic symbolism and sacramentalism – doctrines of incarnation and transubstantiation – were all intensely physical. The relentless symbolization of Catholic life is not just attached to the physical world, it is the physical world' (Haraway, in Goodeve, 2000: 86).

If Haraway's cyborg manifesto was a 'blasphemy' against the grand narratives of redemption within socialism and feminism (1991a: 149) then I am drawn to a 'heretical' reading of her work, which brings to the surface the implicit and buried religious allusions. I dare to suggest that this sense of the *sacramental* achieves precisely what she was intending for her cyborg writing. While eschewing purity, metaphysics and holism, it offers 'the absolute simultaneity of materiality and semiosis' (Goodeve, 2000: 86). A sacramental sensibility speaks

[6] Ironic references to Catholicism, the sacred, Biblical narratives of the Fall abound in her work. See especially Haraway 1991a: 149, 151; 1993; 1997: 131–48.

of the same fusion, of affirming the existence of the cultural and manufactured products of human labour while placing them within a horizon of sacred value which speaks of a transcendental – but not other-worldly – mode of being. Sacraments are thus signs of the 'transfiguration' of the material and not of its effacement or denial. 'We do not encounter God in the displacement of the world we live in, the suspension of our bodily and historical nature' (Williams, 2000: 207). The artefactual world, the realm of *Homo faber* and technologies, as well as that of 'nature', may thus be regarded as a medium of wonder and transfiguration. Those who appeal to the absence of redemption beyond a return to the simplicity of transcendent Being assume that the God who will save us is absent from the modern world.[7] However, a sacramental sensibility 'about the literal nature of metaphor and the physical quality of symbolization' (Goodeve, 2000: 141) attests to the same pattern of materialism and re-enchantment in which transcendence and immanence are intertwined and not separated.

Similarly, in the work of Luce Irigaray, the divine represents such a horizon of incompleteness and becoming – a *telos* – in which all human essences remain unfinished and unfixed. Irigaray's 'divine' beckons creation beyond the ontological hygiene of fixed essences into realizing new, as yet unarticulated possibilities for identity and community. This is not a projection of our interests on to a transcendent realm – the assumption that underpins Haraway's secularism – but serves as an ideal towards which women, especially, can aspire, 'a *sensible transcendental* that comes into being through us' (1993a: 129, emphasis in original), not a God out there against whom mortal beings measure their finitude and imperfection. Irigaray's project is therefore not to retrieve a putative female deity from prehistory, but to engage in the audacious task of imagining a divinity for herself. She relocates 'transcendence' away from other worlds or an immaterial, celestial space, and reconfigures it as 'ever beyond present actuality . . . not reducible to the set of physical particulars of the material universe' (Jantzen, 1998: 271).

While the theological symbolic of Western technoscience is assumed to sanction the disavowal of embodied finitude in the name of a quest for what passes as 'transcendence', alternative genealogies of the relationship between the human, the technological, the natural – and the divine – can be reconceived and put to work. Cyborg writing invites us to confront the fears we have about 'translation' and resist the drift back to the 'purification' of modernity – nostalgia, perfection, closure, dualism and 'the crossed-out God'.

Technologies are important vehicles for human creativity and redemption, but it is necessary to question the assumption that spiritual enlightenment

[7] What if, as Bruno Latour suggests in a rather unkind parody of Heidegger, the gods dwell as much in culture mediated by technology as in the purity of nature, 'in Adidas shoes as well as in the old wooden clogs hollowed out by hand, in agribusiness as well as in timeworn landscapes, in shopkeepers' calculations as well as in Hölderlin's heartrending verse' (Latour, 1993: 66)?

comes at the cost of physicality and corporeality. The effacement of the theological imagination constitutes its repression within the symbolic of modernity, not its destruction. For Irigaray, therefore, new concepts of divinity – as a guarantee against the reification of contingent experience – are fundamental to a renewed ethical and political vision. Commodities and material artefacts are, potentially, redolent with sacred power; and transfiguration and redemption can be achieved within, not beyond, the realms of technologies, human agency and material culture. It is imperative to rethink the model of 'transcendence' that informs representations of the post/human premised on a vision of immortality, omniscience, omnipotence and incorporeality. The task is not simply to interpret the symbolic of transcendence in whose image technoscientific desires for omniscience and necrophilia are legitimated, but to change it.

CHAPTER TEN

Gods and monsters[1]

Humanity is neither an essence nor an end, but a continuous and precarious process of becoming human, a process that entails the inescapable recognition that our humanity is on loan from others, to precisely the extent that we acknowledge it in them . . . Others will tell us if we are human, and what that means. (Davies, 1997: 132)

Two themes have been pre-eminent in this book: the first is the abiding issue of what it means to be human, a question invoked with particular intensity by the proliferation of cybernetic, biomedical and digital technologies over the past fifty years. The second concerns the centrality of narrative – be it scientific, literary or mythical discourses – for supplying Western culture with exemplary and normative representations of what humanity might become in the face of such advanced technoscientific endeavours. In analysing the representations of selected post/human figures – liminal characters, inhabiting the boundary between the human and the almost-human – I have resisted essentialist models of 'human nature', preferring instead to emphasize the way in which definitive versions of what it means to be human emerge from encounters with the refracted 'Other' in the form of the monster, the android, the *Doppelgänger*, or the alien. This chapter returns to these themes and reflects on their ethical, political and theological significance. But first, as much of this book has been about telling stories, a final example from contemporary popular culture will serve to prefigure some crucial issues to do with narrative and post/human identity.

Telling stories, building worlds

The central characters of the film comedy *Galaxy Quest* (1999) are former stars of a long-running television science fiction series, now cancelled from the schedules.

---

[1] Dr Pretorius (to Henry Frankenstein, proposing a toast): 'To a new world of gods and monsters!' (*The Bride of Frankenstein*, 1935). *Gods and Monsters* was also the title of the dramatized story of the life of James Whale, released in 1999 and starring Ian McKellen.

Reduced to making a living from personal appearances (in character) on the science-fiction-fan convention circuit, the actors have become jaded by their own type-casting and thoroughly cynical of the wholesome and humanistic values the space opera once promoted. Their routine is disturbed by the appearance of a species from a distant galaxy, who have mistaken the broadcasts of the series, intercepted from Earth, for a factual account. Lacking the facility to distinguish fact from fiction, the aliens have constructed an entire civilization by mimicking not only the technologies but also the very cultural values of the television series. Under threat from an invading race, the aliens have come to seek the help of the gallant crew, and the actors find themselves expected to embody the very ideals of heroism and honour they had formerly found so unconvincing. They are thrown into a world in which the fictional values of popular entertainment are taken as binding – indeed vital, if the aliens are to withstand their aggressors.

While *Galaxy Quest* plays up the comic effects of this confusion between fact and fiction, at a deeper level the movie's core premise exemplifies one of the key themes of this book, namely the facility of the human creative imagination to fabricate material *and* symbolic worlds. In copying what they see on the television broadcast, the aliens mistake representation – of space travel, advanced technologies, the values of the crew – for 'reality'. Yet they also assume that technological and moral worlds are indivisible, and that there is a fundamental affinity between the physical and the metaphysical.

In similar ways, I have pursued the thesis that scientific and fictional representations of the world share common features. It has not been my intention to debate the plausibility of fictional technology (Krauss, 1996); nor to engage with the 'culture wars' debate about the social construction of science (Ross, 1996a), although I have suggested that scientific representations serve not only to reflect and order reality, but actively to construct it, by facilitating creative activity that constitutes a moral, political – even theological – world. Particular narratives and aspirations are implicit in humanity's engagements with its technologies; and my discussion of the assumptions about exemplary and normative post/humanity underpinning programmes such as AI and the Human Genome Project suggested that tools, machines and artefacts cannot be severed from the values embedded in their design, deployment and distribution (Chapter five). Understandings of technologies as empirical objects, therefore, mere tools in the service of autonomous humanist values, as in the 'instrumental' perspective of technocracy, reviewed in Chapter six and seven, fail to capture just such a 'constitutive' model. Not only does instrumentalism tend to attribute simplistic qualities of homogeneity and determinism to technologies, but it also obscures the extent to which they reflect and further political, cultural and economic values.

I have also been concerned to advance the notion that myth, literature and popular culture are powerful ingredients in constructing world-views which shape political, ethical and technoscientific priorities, such as the visions of unlimited progress and prosperity represented by brands of technocratic futurism (Chapters five, six and seven). This further reinforces a sense that the human imagination – not technoscientific this time, but activities of storytelling and myth-making – is constitutive, a crucial part of building the worlds in which we live. In this context, a 'world' may be composed of material objects, the products of human fabrication; but also comprises signs and symbols which create an environment of meaning which is value-laden and binding (Paden, 1994, 2000). In characterizing (religious) worlds in this fashion, William Paden is influenced by the work of Martin Heidegger, drawing on the latter's characterization of 'being-in-the-world' to construct a materialist, situated model of human creative activity: 'what humans "are" is what they do as agents in their worlds, a world being here a mutually constitutive relation of subject and objects' (Paden, 2000: 337). Heidegger considered technologies to be more than merely tools or techniques – 'the correct [superficial] instrumental definition of technology still does not show us technology's essence' (Heidegger, 1993a: 313) – but as something with a profound potential for constructing human being and becoming.

It follows, therefore, that the question of what it means to be post/human cannot simply be framed as a matter of the degree to which 'technology', be it prostheses, access to cyberspace or genetic modification, be allowed to impinge on 'us'. It is, rather, that the biotechnological, cybernetic and digital age has prompted a recognition that humanity has always co-evolved with, and defined itself in relation to, its environment, tools and technologies. In other words, this is less a matter of the empirical impact of technologies than their significance for human ontology. The impossibility of isolating 'human nature' from its refracted others suggests a model of post/humanity as inextricably bound up in relationality, affinity and contingency.

Technology and ontology

While Heidegger's concern over the technological enframing of Being is often assumed to be articulating a straightforwardly technophobic position, he actually occupies a central role in philosophical articulations of technologies as constitutive (Feenberg, 1991: 4–6). He resists any suggestion that relief from enframing and a restoration of authentic being can come simply from the repudiation of technologies. His strategy is altogether more complex, if problematic. Citing the poet Hölderlin, Heidegger argues that 'where danger is, grows/The saving power also . . .' (1993a: 333), suggesting that redemption from the ossifying powers of technology cannot come from an escape from the technological age but by a 'turning', a willingness to reinvoke the radical openness of Dasein (Bernstein, 1991: 119–120). This assumes the form of a reverential affirmation that is able to outstrip and subordinate what Max Weber would have termed the

223

'iron cage' of technical rationality. If reflective Being can displace enframing as the predominant mode in which ontological concealment unfolds, then humanity's relationships with machines, artefacts and tools can become enhancing. Technology will no longer exert the strait-jacketed absolute claim over existence. Thus freed, humanity can explore anew the mystery and 'giftedness' of Being (Heidegger, 1993a: 338–40; Zimmerman, 1990: 172).

Heidegger regards such a redemptive act as establishing a reinvigorated relationship with Being in which humanity would be disabused of its fantasies of control over technology as a route to authentic existence. Potentially, then, Heidegger calls for a renewed sensibility which allows the natural order an autonomy of its own, liberated from the enframing exploitation of human technological imperatives (Bernstein, 1991: 125). This is not, however, an alternative ordering of technoscientific or socio-economic relations, but a reorientation of *poiésis*. Indeed, in looking for some form of agency by which the enslaving aspects of technology might be averted Heidegger turns not to the person of the scientist or the politician, but to the philosopher, the poet or the artist: 'Philosophy will be unable to effect any immediate change in the current state of the world. This is true not only of philosophy but of all purely human reflection and endeavour. Only a god can save us. The only possibility available to us is that by thinking and poetizing we prepare a readiness for the appearance of a god . . . insofar as in view of the absent god we are in a state of decline (Heidegger, 1976: 277).

In claiming that the only protection from the abyss of *Ge-stell* is higher thinking – 'only a god can save us' – Heidegger therefore opts for an idealist stance at the expense of material, political or economic interventions. Modes of practical engagement – *praxis* or *phronésis* – that privilege the strategic, the value-directed and political as avenues of resistance to the encroachments of enframing, are not considered. His evocation of an idealized deity beyond the sphere of human agency fails to grasp that human being and becoming are always already culturally and materially mediated. While affirming, with Heidegger, the substantive and constitutive character of technologies, therefore, it is necessary also to insist on a model of world-building which places human agency at the heart of cultural and technoscientific activity.

Even though they may become reified and institutionalized, technologies are never purely impersonal, but rooted in productive activity. Thus technologies may be determinative of human experience, but they are not necessarily deterministic. This requires us to abandon the polarities of technophobia and technophilia in favour of a more *reflexive* model. Technologies are the products of human creative activity – as are representational practices – but become in turn the media within which human ontology is realized. If technologies are both the products of human creativity and the environment within which creative labour takes place, then strategies for change need to engage with the vexed question of, as Feenberg puts it, 'who makes technology, why and how' (Feenberg, 1991: 14). As I argued in Chapter five, therefore, it is essential to pay heed to the question of what John Law terms *distribution* – issues of access to technologies, of priority, equity and participation (Law, 1991). The capacity of technologies to participate in the constitutive process of world-building – as possessing the power to shape what it means to be human at a material and a metaphysical level – begs deeper analysis of what powers will shape the digital, cybernetic and biomedical future.

Toward a post/human ethic

Amidst the plethora of possible post/human futures it would be invidious to attempt to reinscribe a single solution to the question of what it means to be human in a digital and biotechnological age. Yet it does seem that some futures are more equal than others, and that it should be possible to articulate some criteria for what might be termed a *post/human ethic*. These engender a political, ethical and theological context within which choices might be made about the various futures represented by the many representations of the post/human. I return, therefore, to themes that have informed my analysis throughout the book so far: representation, monstrosity and *alterity*, the contingency of human identity and the resurgence of the sacred.

REPRESENTATION

The invocational genie is already out of the bottle. There is no retreat to innocence. We are already cyborgs . . . The questions that remain are political, and start with paying attention to the invocational voice. Who can speak? Who is silenced? Who is commanded? Who inscribes digital domains? Who writes the spells? (Chesher, 1997: 91)

My critique of some of the most prominent representations of what it means to be post/human has revealed a multiplicity of visions and political/economic interests. I have been particularly critical of models of post/humanity which adopt self-fulfilling prophecies of evolution, either to foretell human obsolescence or to predict an era of the superhuman. Deterministic accounts of this process, either those that invoke such abstractions as 'the gene', or depict human development as an impersonal drive toward disembodied rationality, are a surrender of human reflexivity, or the power to act creatively: to be authors as well as subjects of history. I have also exposed narratives of post/human futures and technoscientific development which effectively remake the future in the image of privileged minorities while masquerading as universal philosophies of mind or other reductionistic models of human nature. The very ubiquity of hybrid, fabricated, enhanced and cyborged humanity – not to mention the blurring of nature and culture that entails – means that it is impossible to seek refuge in essentialist or nostalgic appeals to human nature hedged around by ontological hygiene.

In beginning to consider the prospects for critical interventions in relation to the future directions of technoscience, critical attention to the terminology of *representation* has been invaluable. It has extended beyond an association with depiction and portrayal, to encompass that which *stands* for another, not so much in a linguistic or semiotic sense as in terms of that which is the surrogate or the exemplar of something else. As I suggested in Chapter five, for example, work on mapping the human genome engenders a 'code of codes', a bio-informatic database which purports to be the distillation of humanity's genetic composition. 'The Genome' has effectively displaced the human individual from the discourse of normative humanity, and is transformed into the consummation of 'human nature' in all its diversity. Whatever the representation once stood for or simulated has been completely displaced by the technoscientific artefact itself. After Jean Baudrillard, we might imagine that the representation (in the form of the gene/genome) has become the simulacrum of the actual human, displacing it to the point where the reproduction/ representation has effaced the original (Baudrillard, 1983).

Representation also suggests a link to a more formal political domain. To ask how something or somebody is *represented* opens questions about power and authority, raising questions of who speaks for whom, as a matter of advocacy. Scientists may claim that the encoded genome represents universal humanity, but concern is mounting that it is in fact a partial representation that will occlude vital ethnic diversity (Hubbard and Wald, 1997; Haraway, 1997). So who is appointed to speak on whose behalf? Depictions of technologized humanity that articulate visions of post/human futures may only be 'representative' of certain privileged groups, reflecting the wish-fulfilments of some people who assume the right to speak on the basis of extrapolations of their own (partial and self-interested) perspective. Representations that are ideological or reductionist – humans as genes, machines, nature as feminized other – serve to enshrine and reify certain assumptions about normative or exemplary humanity, but at the expense of excluding others from the discourse altogether. Chapter seven considered what happens when questions of equity, distribution and non-representation – at a political and economic level – go unaddressed. This is one dimension of a political dimension to genealogical critique, first explored in Chapter two: what human beings take for granted about themselves is contingent, and rests on repression and non-representation, be that linguistic, political or economic. The taxonomical purity of immutable, definitive 'human nature' cannot help but evoke the very almost-human beings whom it seeks to repress in the name of ontological hygiene, those – human, non-human and almost-human – who are thereby objectified or excluded. As Bruno Latour hints, therefore, 'representation' is a matter of how the world is ordered, not only in terms of taxonomy, but also in terms of *governance* (Latour, 1993).

MONSTROSITY, GENEALOGY, ALTERITY

Perhaps it is time to ask the question that always arises when the monster is discussed seriously . . . Do monsters really exist? Surely they must, for if they did not, how could we? (Cohen, 1996b: 12)

In concentrating on the notion of monstrosity in this book, I have tried to pursue a number of ideas. Historically, the mingling of wonder and fear occasioned by such fabulous creatures served an important function as a vehicle of moral and theological instruction. Yet it is clear that the fascination with such creatures – a fascination that spanned contemporary science fiction – was to do with their ability to refract and delineate prevailing orthodoxies about what it meant to be human. Those on the boundaries were gatekeepers between identity and difference – gendered, racialized, medicalized. Teratology functioned to show forth alterity, the objectified other, as well as to disrupt and thus to question the axioms of identity. Teratology gave didactic prominence to creatures of fear and hope, reminding its audience of the dissonant, the alien and the marginal whose indeterminate status simultaneously made possible and subverted fixed essences, secure identities and the distinctiveness of species. Yet this very visibility was a reminder of the strategies cultures use to conceal the hybridity and translation that is everywhere and which always carries the suggestion that 'we are all monsters, outrageous and heterogenous collages' (Law, 1991: 18).

Foucault's genealogical critique, similarly, regarded the tracing of essential and ahistorical categories or deterministic teleologies as impossible and irrelevant. What mattered was the excavation of the dynamics of power and difference that characterize any given discipline of knowledge or regime of governance. Even though this shifts the emphasis from ontology to ethics and politics, it is possible to think about exclusion, difference and equity on those terms. At the heart of fabulation, teratology and genealogy lies the cultivation of the unfamiliar for critical and ethical effect. This should encourage interpreters of representations of the post/human to be mindful of the invisibility or objectification – the misrepresentation – of those whose existence guarantees coherent categories, but whose non-participation or exclusion underpins the prosperity and security of others. I argued in Chapter six that while ostensibly, Star Trek is a narrative of equality, diversity and tolerance, it prefers the options of conformity and assimilation to the incorporation of radical post/human difference. Star Trek may appear progressive, but in its anxieties and depictions of alterity and monstrosity it reveals a fear of technological encroachment upon the sovereign male rational subject (assumed to be representative of all humanity) – especially upon his body. Similarly, the ontological status of Frankenstein's creature and the golem can be read as highly contingent, dependent on the assertion of normative humanity against which such almost-human beings are constructed not as autonomous agents but as refracted 'Others' (Chapters three and four).

227

Thus, a teratological analysis suggests that any attempt by humanity to delineate an absolute 'human nature' as a form of ontological purity cannot fail, paradoxically, to evoke its others, thereby subverting its own stability and fixity. In distinguishing human nature from the almost-human, discourses are composed of *différance* – a definition that establishes both difference and deferral. This is particularly reminiscent of Jacques Derrida's critical method, which emphasizes the degree to which all forms of language, speech, signification are *textual* in nature and caught in the system of referentiality. This method, which Derrida terms 'grammatology', or the science of writing, exposes the illusions of the metaphysics of presence, and is described by him, intriguingly, as 'beyond the closure of knowledge . . . a sort of monstrosity' (Derrida, 1978: 4–5; see also 167–8).

CONTINGENCY AND CO-EVOLUTION

I have argued that the digital, cybernetic and biotechnological age occasions a questioning of boundaries and reordering of human relationships with non-human nature and technologies. The image of the pristine humanist subject untouched and unpenetrated by invasive technologies, the hyper-macho invulnerable techno-hero or transhumanist sublime, all suffer from what I called an 'ontological hygiene' with respect to human nature. This sort of thinking is eschewed by contemporary thinkers such as Bruno Latour and Donna Haraway, both of whom start from the premise that in a highly technologized world, whatever 'human nature' may be, it is hybridized, more to do with boundaries than essences. I have interpreted the performance art of Stelarc, reviewed in Chapter eight, as a practical example of such a perspective. For Stelarc, Latour and Haraway, we have always been mixed up, co-evolving with our tools, living a hybrid existence. The rapid intensification of new technologies over the past fifty years may have accentuated this realization; but there are important continuities with models of prehistoric human evolution which emphasize similarly contingent, reflexive models of cultural development (Tattersall, 1998; Haraway, 1992b).

To think of humans as 'cyborgs' (as in the work of Donna Haraway, discussed in Chapter nine), is to conceive of post/humanity as divested of the illusions of innocence, purity, detachment and the pretence of transcendence – all dangerous facets of a kind of will to power. The cyborg delights in her technological complicity and recognizes the plasticity of categories of being. Yet as Chapter nine argued, Haraway's vision is problematic in that she chooses to reinvent the cyborg in ways that underestimate the enduring power of the *Terminator-Robocop*-style omnipotent killing machine of science fiction and which undervalues a post-secular critique of her particular brand of Marxist-feminism. Despite this, however, the cyborg is a handy metaphor for post/human experiences which defy models of technology as a deterministic monolithic force or as a 'quick fix' whose social and political implications are somebody else's concern. Global technoscience reduces everything into an artefact, a thing made not born, while representing nature as primal, innocent and independent of human agency. But what we call 'nature' is already heavily managed by technoscientific interests, such that any notion of 'nature' outside 'culture' 'is not so much elsewhere as nowhere' (Haraway, 1992a: 295) – in other words, it does not exist. It would therefore be inappropriate to build an ethic on an imagined organic unity with such a nostalgic construction, but the challenge is to express new forms of relationality that embody affinity and difference but not dominion. Ethically and experientially, the cyborg is a heuristic figure that suggests the rejection of solutions of either denial or mastery in favour of a post/human ethic grounded in complicity with, not mastery over, non-human nature, animals and machines.

The sense that humans and machines are increasingly assimilated, that human nature cannot be realized apart from its tools and artefacts (either as objects of fear or as instruments of mastery), is thus a more authentic understanding of post/human ontology in a digital and biotechnological age. It is also a profoundly materialist understanding, because it refuses to believe either that humanity can retreat to some pure unadulterated 'human nature' independent of the worlds it makes; or that technologies can be exploited to transcend bodily finitude and limitation. That 'humanity' is actually constituted in reflexive interaction – even co-evolution – with tools, environment and artefacts is a suggestion I advanced at the end of Chapter eight and is an extrapolation of the 'constitutive' model of technologies.

The ineffability and autonomy of technologies is also at the heart of Asimov's robot stories (see Chapter eight) and develops the evocative suggestion that neither nature nor artefact will ever be entirely within human control. Chapter four, similarly, traces the many histories of the golem as demiurgical device and potential sentient being. Marge Piercy's novel poses the question of whether Yod the cyborg/golem could retain his purpose as an object of another's bidding and ever attain genuinely free personhood. At the root of Frankensteinian fears of 'creation out of control' is an expectation, rarely challenged, that non-human nature, tools and artefacts may not be completely within human mastery. However, I argued that Victor's culpability rested in his disregard for precepts of *responsible knowing*, a model that would be capable of acknowledging humanity's role in the fabrication of nature and technoscience, while realizing the autonomy and irreducibility of human tools, artefacts and technologies.

It may thus be necessary to consider the *otherness* of artefacts and the necessary non-consanguinity between humans and things. The (Freudian) claim that machines are never more than human prostheses (Mazlish, 1993: 228) – vestiges of instrumentalism again – does nothing for conceptions of non-human animals or even artefacts as independent, autonomous or ineffable. Mazlish's logic of the inevitability of human continuity – with tools and artefacts, with non-human nature, with the cosmos – honours neither the novelty and distinctiveness of human intelligence (as opposed, even, to that of higher primates) nor facilitates the autonomy of what Donna Haraway terms the 'inappropriate/d Others' of nature and technology (Haraway, 1992a: 295). Notions of affinity and distinctiveness, autonomy and co-evolution (that humans, nature and technologies are mutually constitutive of one another) may therefore better safeguard the moral integrity of humanity's others as well as curbing the hubristic drive of secular humanism's self-absorption.

RE-ENCHANTMENT

I have argued that many representations of the post/human are subtended by discourses of 'transcendence', as equated with idealism and dualism, of the physical world as an encumbrance and illusion, an implicit denigration of embodiment and materialism. Such an understanding of human nature has in the past been used as a rationalization for dominion over the non-human, part of a presumption that to be godlike is to seek mastery over creation, heedless of the fragility and interdependence of life. I have challenged the assumption that human inventiveness could be adequately characterized as a universal drive to abandon bodily finitude, to cheat death and abandon the sensual world, even though this instinct has been rationalized as an innate religious impulse to 'play God' or to become like gods.

Once again, the inadequacy of the polarization of technophobia and technophilia resurfaces. It is commonplace to regard instrumental technocracy as unambiguously technophilic, valorizing the unlimited capabilities of human inventiveness. Yet in my characterization of Victor Frankenstein's confusion about the boundaries between life and death (Chapter three) and in the parallels between Grace Jantzen's critique of Western necrophilia and the symbolic system of 'transcendence' in much of transhumanism (Chapter seven), I suspect that such uncritical embrace of technological omnipotence, omniscience and immortality betrays not so much a love of life as, paradoxically, a pathological fear of death, vulnerability and finitude. Aspirations toward a digitalized post-biological humanity often reflect the desire for a spiritualized, non-corporeal body as the fulfilment of a disdain for the mortality of the flesh. Technophobia, ostensibly, seems to be driven by denial of progress, fear of change, loss of control; but the evidence suggests that so-called technophilic attitudes are subtended by similar projections, valorizing technologies as protections against fears of vulnerability, contingency, impurity and mortality.

Religion, culture and gender

Much of the fascination with virtual technologies considered in Chapter seven represents the re-enchantment of a world thought to have been denuded of religion. Such 'technochantment' shares with ancient world-views such as Gnosticism an enduring fascination with a transport into a hidden, spiritual realm of celestial wisdom beyond the base contingency of the material world. '[W]hat attracts us in the fleshly world is no more than an outer projection of ideas we can find within us' (Heim, 1993a: 64). Erik Davis's analysis of 'techgnosis' manifested further reverberations of idealism and dualism, of regarding the known, immanent world as illusory, ephemeral and (literally) immaterial, serving as a rationalization for the 'mastery' of non-human nature and gendered attitudes to knowledge and embodiment. Such visions also bear similar traits to hermetism, in its understanding of reality as information, encoded in material form but essentially immaterial in nature; and in metaphors of technoscientific endeavour as a quest to impose human domination over nature. It serves as the ideological underpinning for many of the dualisms that have fuelled modernity and technoscientific innovation. The predilection for the qualities of detachment, omniscience, immutability and incorporeality subtend particular ways of knowing in classical artificial intelligence; promote disdain for the embodied contingency and foster technologies that are obsessed with cheating death, vulnerability and finitude.

This is another reiteration of ontological hygiene, an assertion that human destiny lies in the pursuit of a fantasy of 'transcendence' – the emulation of the 'crossed out God' of modernity (Latour, 1993: 128) – in the pursuit of an omnipotent, incorporeal, immortal, invulnerable route of becoming. In Latour's terms, it is a further exemplification of modernity's attempts to impose a taxonomic 'purification' on the world, by segregating matter and spirit, secular and sacred. Yet the boundary between material and spiritual is as much a construct as that between humans, animals and machines. As I argued in Chapter nine, to interpret 'transcendence' – and, by implication, any evocation of religion and the sacred – as inevitably implying the worship of incorporeality, immortality and immateriality is to collude with its ideological usage, and to fail to analyse the very constructedness of such a notion. To equate a drive for transcendence with an abandonment of physical worlds is, therefore, to extrapolate a particular religious symbolic system peculiar to Western modernity – which represents only specific social, gendered and cultural interests – into a universal human essence.

Beyond 'transcendence'

Those who suggest a continuing affinity between Western technoscience and religious impulses are therefore implicitly colluding with a number of assumptions about religion, culture and gender. They assume that there is an innate drive towards disembodied transcendence deeply embedded in every human psyche, regardless of gender, racial or religious background. Furthermore, a spiritual quest is necessarily seen as the search for disembodied omnipresence of a kind discerned in the supposedly Platonic forms of pure information and perfect mastery. However, all this talk of 'cyberspace as sacred space' (Chapter seven) cannot assume that the religious revivals and preoccupations of twentieth and twenty-first-century late capitalism are a particularly reliable indicator of universal human spiritual motivations, or that 'technochantment' is the only mode of discerning transcendence in the material and technological world.

Given that these notions of 'playing God' or 'becoming like gods' still retain such a powerful hold on the Western imagination, it is not sufficient to return to secular critiques of religion, as David Noble or Donna Haraway have done. It is doubtful whether Noble's solution, of a secularized 'successor' science, free of hubristic and imperialistic impulses, for example, would serve up a more equitable model of scientific practice and technological development. Rather, an alternative religious symbolic system needs to be articulated in order for different kinds of relationships to the material world to be enacted, ones which value different kinds of scientific epistemologies and technological endeavours. If Frances Yates's interpretation of hermetism is correct, then technochantment, like all models of technoscientific endeavour, comprises a number of heterogenous elements: not only those that valorize the flight from the material world in pursuit of ultimate knowledge and power; but also those that celebrate the potential of immanence and embodiment, and seek to sacralize the known world and infuse material experience with the possibilities of divinity. These represent wholly different relationships with the material world (including birth, embodiment and mortality), a greater tolerance for human contingency and non-human nature – and potentially, an alternative symbolic system of religion, culture and gender. 'Our longings are not disconnected from our present "bodily-living-in-the-world" and to recognise this is to bring "transcendence" back into the sphere of "immanence"' (Carrette and King, 1998: 141).

My dissatisfaction with assertions of ontological hygiene, therefore, has been in the name of the principle of the interrelatedness of things, the vitality of matter and the irreducibility of the other. The displacement of the taken-for-grantedness of ontological hygiene, the presence of the monstrous and the traces of affinity between humans, nature and technologies all represent a repudiation of the world as determined, inert, evacuated of the divine or reduced to the image of anthropocentric desires. As I argued in Chapter eight, cybernetics, too, suggests that all of life is in process, without functional limit, but forever in the midst of movement and interchange. Another way of putting this would be to say, after Bruno Latour, that the world is too full of vitality to surrender to our attempts at purification. There is forever new information to be decoded, and the world is both available for modification through intentional activity and always already exceeding the grasp. 'The System is not closed; the sacred image of the same is not coming. The world is not full' (Haraway, 1992a: 327).

It is beyond the scope of this book to offer more than a critique of the ideology of transcendence as invoked in representations of the post/human. Clearly, a future task remains to embody more fully a reconstruction of appeals to 'transcendence', but in contrast to the kinds of technotranscendence I outlined earlier, such an apprehension of the sacred and divine would be expressed through the medium of embodied, contingent experience – as in the 'sacramental' perspective indicated in Chapter nine. This would acknowledge the fabricated, technologized worlds of human labour and artifice as equally capable of revealing the sacred as is the innocence of 'nature'. To conceive of alternative notions of 'transcendence' is possible, and in very different ways from the incorporeal flight from materiality and finitude so often equated with religion. What is more, it has, potentially, a potential to effect a critical and transformative sensibility. The human, the material and the semiotic are inseparable; but out of that there arises a novelty, what Irigaray terms a 'horizon' (Irigaray, 1987: 66–9) of new ways of being. This holds out a number of possibilities for thinking about human engagement with technologies and with humanity's own creative potential (especially to create and enhance life and to transform our material surroundings with beauty and utility), and of continuing to consider how that engagement with the material might be an avenue into 'transcendence' or divinity. One of my criticisms of goddess spirituality has been the risk that it locates the sacred exclusively in a fantasy of unadulterated 'nature', and withholds the possibilities of its being experienced in the fabricated world of technological culture. Yet if nature, and fantastic creations such as myths and monsters, can evoke a sense of awe and wonder, why not technologies?

How could we be capable of disenchanting the world, when every day our laboratories and our factories populate the world with hundreds of hybrids stranger than those of the day before? . . . How could we be chilled by the cold breath of the sciences, when the sciences are hot and fragile, human and controversial, full of . . . subjects who are themselves inhabited by things? (Latour, 1993: 115)

In many respects, therefore, it is legitimate to regard the emergent cybernetic, biotechnological and digital age as representing a new era of post/human history. Yet in other ways, the contemporary West retains a marked degree of continuity with more ancient cultures that first dreamed of gods and monsters at the margins of human imagining. An exploration of the many different hopes and fears surrounding the impact of advanced technologies requires, as I have been arguing, a full register of representational practices, cultural, literary, mythical and scientific. Monsters, aliens and others still function as important monitors and mediators of understandings of what it means to be post/human, not least in their indeterminacy, their eschewal of ontological purity, and their attention to human nature as defined by boundaries rather than essences. They embody the disturbing reminders of difference at the heart of unitary identity, and they suggest that any post/human ethic can be neither an escape into technocratic invulnerability nor a retreat into the imagined purity of organic essentialism. Rather, as I have argued, it will be about the pleasures and risks of multiple allegiances, contingent identities and nomadic sensibilities. Fantastic encounters with representations of the post/human offer important insights into the many meanings of being human, but they are also devices by which new worlds can be imagined.

REFERENCES

Ackroyd, Peter (1993), *The House of Doctor Dee*, London, Hamish Hamilton.

Adam, Alison (1998), *Artificial Knowing: Gender and the Thinking Machine*, London, Routledge.

Aeschylus [c. 463 BCE] (1961), *Prometheus Bound*, trans. and intro. P. Vellacott, London, Penguin.

Agnew, Bruce (1999), 'Will we Be One Nation, Indivisible?' *Scientific American*, 10:3, 76–9.

Aldiss, Brian (1973), *Billion Year Spree: The History of Science Fiction*, London, Weidenfeld & Nicolson.

Alexander, David (1996), 'Terrance Sweeney's "God and Roddenberry"', in D. Alexander (ed.), *StarTrek Creator: The Authorized Biography of Gene Roddenberry*, London, Boxtree, 567–75.

Alexander, Philip (1992), 'Pre-Emptive Exegesis: Genesis Rabba's Reading of the Story of Creation', *Journal of Jewish Studies*, 43:2, 230–45.

Allen, Glen Scott (1992), 'Master Mechanics and Evil Wizards: Science and the American Imagination from Frankenstein to Sputnik', *Massachusetts Review*, 33, 505–58.

Alsford, Mike (2000), *What If? Religious Themes in Science Fiction*, London, Darton, Longman and Todd.

Arac, Jonathan (1988), Introduction in J. Arac (ed.), *After Foucault: Humanistic Knowledge, Postmodern Challenges*, New Brunswick, N.J., Rutgers University Press, vii–x.

Argyle, Katie and Rob Shields (1996), 'Is there a Body in the Net?' in R. Shields (ed.), *Cultures of Internet: Virtual Spaces, Real Histories, Living Bodies*, London, Sage, 58–69.

Aronowitz, Stanley, Barbara Martinsons and Michael Menser (eds) (1996), *Technoscience and Cyberculture*, London, Routledge.

Asimov, Isaac [1950] (1995a), 'The Evitable Conflict', in *The Complete Robot*, London, HarperCollins, 546–74.

—— [1976] (1995b), 'The Bicentennial Man', in *The Complete Robot*, London, HarperCollins, 635–82.

Atzori, Paolo and Kirk Woolford (1995), 'Extended-Body: Interview with Stelarc', *Ctheory* (online), available at http://www.ctheory.com/a29extended_body.html [accessed 14 July 1999].

Baldick, Chris (1987), *In Frankenstein's Shadow: Myth, Monstrosity and Nineteenth Century Writing*, Oxford, Oxford University Press.

—— (1995), 'The Politics of Monstrosity', in F. Botting (ed.), *Frankenstein: Contemporary Critical Essays*, London, Macmillan, 48–67.

Balsamo, Ann (1995), 'Forms of Technological Embodiment: Reading the Body in Contemporary Culture', *Body & Society*, 1:3–4, 215–37.

—— (1996), *Technologies of the Gendered Body: Reading Cyborg Women*, Durham, NC, Duke University Press.

Bann, S. (1994), Introduction in S. Bann (ed.), *Frankenstein, Creation and Monstrosity*, London, Reaktion Books, 1–15.

Barbour, Ian (1993), *Ethics in an Age of Technology*, London, SCM Press.

Barr, Marleen S. (1992), *Feminist Fabulations: Space/Postmodern Fictions*, Ames, Ind. Iowa State University Press.

—— (1995), '"All Good Things . . ." The End of Star Trek: The Next Generation, The End of Camelot – The End of the Tale about Woman as Handmaid to Patriarchy as Superman', in T. Harrison, S. Projansky, K.A. Ono and E.R. Helford (eds), *Enterprise Zones: Critical Positions on Star Trek*, Boulder, CO., Westview Press, 231–43.

Barrett, Michèle (1980), *Women's Oppression Today: Problems in Marxist Feminist Analysis*, London, Verso.

—— and Duncan Barrett (2001), *Star Trek: The Human Frontier*, Cambridge, Polity Press.

Baudrillard, Jean (1983), *Simulations*, trans. P. Foss, P. Patton and P. Beitchman, New York, Sémiotexte.

—— (1988), *Selected Writings*, ed. Mark Poster, Oxford, Polity Press.

Bauman, Zygmunt (1998), 'Parvenu and Pariah: Heroes and Victims of Modernity', in J. Good and I. Velody (eds), *The Politics of Postmodernity*, Cambridge, Cambridge University Press, 23–35.

Beer, Gillian (1970), *The Critical Idiom: The Romance*, London, Methuen.

Benedikt, Michael (1993), Introduction in M. Benedikt (ed.), *Cyberspace: First Steps*, Cambridge, MA, MIT Press, 1–25.

Benjamin, Walter [1936] (1992), 'The Work of Art in the Age of Mechanical Reproduction', in *Illuminations*, ed. and intro. Hannah Arendt, trans. H. Zohn, London, Fontana, 211–44.

Bennett, Jane (1997), 'The Enchanted World of Modernity: Paracelsus, Kant, and Deleuze', *Cultural Studies*, 1:1, 1–28.

Bennington, Geoffrey (1998), 'Derrida', in S. Critchley and W.R. Schroeder (eds), *A Companion to Continental Philosophy*, Oxford, Blackwell, 549–58.

Bernauer, James (1999), 'Cry of Spirit', in J.R. Carrette (ed.), *Religion and Culture by Michel Foucault*, Manchester, Manchester University Press, xi–xvii.

Bernstein, Richard (1991), *The New Constellation: The Ethical-Political Horizons of Modernity/Postmodernity*, Cambridge, Polity Press.

Bey, Hakim (1998), 'The Information War', in J. Broadhurst Dixon and E.J. Cassidy (eds), *Virtual Futures: Cyberotics, Technology and Post-Human Pragmatism*, London, Routledge, 3–8.

Bilski, Emily D. (1988), 'The Art of the Golem', in E.D. Bilski (ed.), *Golem! Danger, Deliverance and Art*, New York, New York Jewish Museum, 44–111.

Bilski, Emily D. and Moshe Idel (1988), 'The Golem: An Historical Overview', in E.D. Bilski (ed.), *Golem! Danger, Deliverance and Art*, New York, New York Jewish Museum, 10–14.

Boeke, Richard (1997), 'The Theology of Star Trek', *Faith and Freedom*, spring/summer, 48–53.

Bolter, J.D. (1984), *Turing's Man: Western Culture in the Computer Age*, London, Duckworth.

Borgmann, A. (1984), *Technology and the Character of Contemporary Life*, Chicago, University of Chicago Press.

Bostrom, N. (ed.) (1999), 'The Transhumanist FAQ' (online), available at htttp://www.transhumanist.org [accessed 20 August 2000].

Botting, Frank (1991), *Making Monstrous: Frankenstein, Criticism, Theory*, Manchester, Manchester University Press.

—— (ed.) (1995), *Frankenstein: Contemporary Critical Essays*, London, Macmillan.

—— (1999), *Sex, Machines and Navels*, Manchester, Manchester University Press.

Boyd, Katrina G. (1996), 'Cyborgs in Utopia: The Problem of Radical Difference in *Star Trek: The Next Generation*', in T. Harrison, S. Projansky, K.A. Ono and E.R. Helford (eds), *Enterprise Zones: Critical Positions on Star Trek*, Boulder, CO., Westview Press, 95–113.

Brahm, Gabriel (1995), 'The Politics of Immortality: Cybernetic Science/Fiction and Death', in G. Brahm, Jr. and Mark Driscoll (eds), *Prosthetic Territories: Politics and Hypertechnologies*, Boulder, CO., Westview Press, 94–111.

Braidotti, Rosi (1991), *Patterns of Dissonance: A Study of Women in Contemporary Philosophy*, Cambridge, Polity Press.

—— (1994a), 'Mothers, Monsters, and Machines', in *Nomadic Subjects: Embodiment and Sexual Difference in Contemporary Feminist Theory*, New York, Columbia University Press, 75–94.

—— (1994b) 'What's Wrong with Gender?' in F. van Dijk Hemmes and A. Brenner (eds), *Reflections on Theology and Gender*, Kampen, Netherlands, Kok Pharos, 49–70.

—— (1996a), 'Signs of Wonder and Traces of Doubt: On Teratology and Embodied Differences', in N. Lykke and R. Braidotti (eds), *Between Monsters, Goddesses and Cyborgs: Feminist Confrontations with Science, Medicine and Cyberspace*, London, Zed Books, 135–52.

—— (1996b), 'Cyberfeminism with a Difference' (online), available at http://www.let.rud.hi/womens_studies/rosi/cyberfem.htm [accessed 4 April 1999].

Brasher, Brenda E. (1996), 'Thoughts on the Status of the Cyborg: On Technological Socialization and its Link to the Religious Function of Popular Culture', *Journal of the American Academy of Religion*, 59:4, 809–30.

Bromberg, Heather (1996), 'Are MUDs Communities? Identity, Belonging and Consciousness in Virtual Worlds', in R. Shields (ed.), *Cultures of Internet: Virtual Spaces, Real Histories, Living Bodies*, London, Sage, 143–52.

Brooks, Peter (1995), 'What Is a Monster?' in F. Botting (ed.), *Frankenstein: Contemporary Critical Essays*, London, Macmillan, 81–106.

Brown, Kathryn S. (1999), 'Smart Stuff', *Scientific American*, 10:3, 72–3.

Bruce, Donald and Ann Bruce (eds) (1998), *Engineering Genesis: The Ethics of Genetic Engineering in Non-Human Species*, London, Earthscan.

Buchanan, Ian (2000), 'Introduction – Other People – Ethnography and Social Practice', in G. Ward (ed.), *The Certeau Reader*, Oxford, Blackwell, 97–100.

Bukatman, Scott (1993), *Terminal Identity: The Virtual Subject in Postmodern Science Fiction*, Durham, NC, Duke University Press.

Burke, Edmund [1790] (1983), *Reflections on the Revolution in France*, ed. and intro. Conor Cruise O'Brien, London, Penguin.

Butler, Judith (1993), 'Contingent Foundations: Feminism and the Question of "Postmodernism" ', in J. Butler and J.W. Scott (eds), *Feminists Theorize the Political*, London, Routledge, 3–21.

—— [1990] (2000), *Gender Trouble: Feminism and the Subversion of Identity*, 2nd edn, London, Routledge.

Butler, Marilyn (1998), Introduction in Mary Shelley, *Frankenstein, or the Modern Prometheus*, Oxford, Oxford University Press, ix–li.

Canguilheim, Georges (1989), *The Normal and the Pathological*, trans. C.R. Fawcett, New York, Zone.

—— (1992), 'Machine and Organism', in J. Crary and S. Kwinter (eds), *Incorporations*, London, Zone Books, 45–69.

Cantor, Charles (1992), 'The Challenges to Technology and Informatics', in D. Kevles and L. Hood (eds), *The Code of Codes: Scientific and Social Issues in the Human Genome Project*, Cambridge, MA, Harvard University Press, 98–111.

Carrette, Jeremy R. (1999), 'Prologue to a Confession of the Flesh', in J.R. Carrette (ed.), *Religion and Culture by Michel Foucault*, Manchester, Manchester University Press, 1–47.

Carrette, Jeremy R. and Richard King (1998), 'Giving "Birth" to Theory: Critical Perspectives on Religion and the Body', *Scottish Journal of Religious Studies*, special edn, 'Beginning with Birth', 19:1, 123–43.

Carter, Angela (1982), 'Nothing Sacred: Selected Writings', London, Virago.

Casper, Monica J. (1995), 'Fetal Cyborgs and Technomoms on the Reproductive Frontier: Which Way to the Carnival?' in C.H. Gray (ed.), *The Cyborg Handbook*, London, Routledge, 183–202.

de Certeau, Michel [1987] (2000), 'Walking in the City', in G. Ward (ed.), *The Certeau Reader*, Oxford, Blackwell, 101–18.

Chasin, Alexandra (1995), 'Class and its Close Relations: Identities among Women, Servants, and Machines', in J.M. Halberstam and I. Livingston (eds), *Posthuman Bodies*, Bloomington, Indiana University Press, 73–96.

Cherny, Lynn and Elizabeth Weise (eds) (1996), *Wired Women: Gender and New Reality in Cyberspace*, Seattle, WA, Seal Press.

Chesher, Chris (1997), 'The Ontology of Digital Domains', in D. Holmes (ed.), *Virtual Politics: Identity and Community in Cyberspace*, London, Sage, 79–92.

Christ, C.P. (1979), 'Why Women Need the Goddess: Phenomenological, Psychological, and Political Reflections', in C.P. Christ and J. Plaskow (eds), *Womanspirit Rising: A Feminist Reader in Religion*, San Francisco, Harper & Row, 273–87.

—— (1989), 'Rethinking Theology and Nature', in C.P. Christ and J. Plaskow (eds), *Weaving the Visions: New Patterns in Feminist Spirituality*, San Francisco, Harper & Row, 314–25.

—— (1997), *Rebirth of the Goddess: Finding Meaning in Feminist Spirituality*, Reading, MA, Addison-Wesley.

Clark, Stephen R.L. (1995), *How to Live Forever: Science Fiction and Philosophy*, London, Routledge.

Clynes, M.E. and N. Kline [1960] (1995), 'Cyborgs and Space', in C.H. Gray (ed.), *The Cyborg Handbook*, London, Routledge, 29–34.

Cohen, Jeffrey Jerome (1996a), 'In a Time of Monsters', in J.J. Cohen (ed.), *Monster Theory: Reading Culture*, Minneapolis, University of Minnesota Press, vii–xiii.

—— (1996b), 'Monster Culture (Seven Theses)', in J.J. Cohen (ed.), *Monster Theory: Reading Culture*, Minneapolis, University of Minnesota Press, 3–25.

Cole-Turner, Ronald (1993), *The New Genesis: Theology and the Genetic Revolution*, Louisville, Ky., Westminster/John Knox Press.

Collins, Francis S. et al. (1998), 'New Goals for the U.S. Human Genome Project: 1998–2003', *Science* (online), available at http://www.sciencemag.org/cgi/content/full/282/5389/682/html [accessed 6 February 2000].

Collins, Steven F. (1996), '"For the Greater Good": Trilateralism and Hegemony in *Star Trek: The Next Generation*', in T. Harrison, S. Projansky, K.A. Ono and E.R. Helford (eds), *Enterprise Zones: Critical Positions on Star Trek*, Boulder, CO., Westview Press, 137–56.

Cool, Jenny (1998), 'Of Tools and Toys: Donna Haraway's Cyborgs and the Power of Serious Play' (online), available at http://www.sirius.com/~ovid/coolonharaway.html [accessed 11 June 1999].

Cooper, David E. (1995), 'Technology: Liberation or Enslavement?' in R. Fellows (ed.), *Philosophy and Technology*, Cambridge, Cambridge University Press, 7–18.

References

Cowan, Ruth Schwartz (1992), 'Genetic Technology and Reproductive Choice: An Ethics for Autonomy', in D. Kevles and L. Hood (eds), *The Code of Codes: Scientific and Social Issues in the Human Genome Project*, Cambridge, MA, Harvard University Press, 244–63.

Crewe, Jonathan (1997), 'Transcoding the World: Haraway's Postmodernism', *Signs: Journal of Women in Culture and Society*, 22:4, 891–905.

Curran, Andrew and Patrick Graille (1997), 'The Faces of Eighteenth-Century Monstrosity', *Eighteenth-Century Life*, 21:2, 1–15.

Dan, Joseph (1998), 'The Kabbalah of Johannes Reuchlin and its Historical Significance', in J. Dan (ed.), *The Christian Kabbalah*, Cambridge, MA, Harvard University Press, 55–95.

Darnovsky, M. (1991), 'Overhauling the Meaning Machines: An Interview with Donna Haraway', *Socialist Review*, 21:2, 65–84.

Davidson, Arnold I. (1986), 'Archaeology, Genealogy, Ethics', in D.C. Hoy (ed.), *Foucault: A Critical Reader*, Oxford, Blackwell, 221–34.

—— (1991), 'The Horror of Monsters', in J.J. Sheehan and M. Sosna (eds), *The Boundaries of Humanity: Humans, Animals, Machines*, Berkeley, University of California Press, 36–67.

Davies, Tony (1997), *Humanism*, London, Routledge.

Davis, Erik (1993), 'Techgnosis: Magic, Memory, and the Angels of Information', *South Atlantic Quarterly*, 92:4, 585–616.

—— (1995), 'Technopagans', *Wired Magazine*, July (online), available at http://www.wired.com/archive/3.07/technopagans/pr.html [accessed 18 March 1999].

—— (1996), 'Technoculture and the Religious Imagination' (online), available at http://www.levity.com/figment/technoculture/html [accessed 9 March 1999].

—— (1999), *Techgnosis: Myth, Magic and Mysticism in the Age of Information*, London, Serpent's Tail.

Dawkins, Richard (1976), *The Selfish Gene*, Oxford, Oxford University Press.

—— (1998), 'What's Wrong with Cloning?' in M.C. Nussbaum and C.R. Sunstein (eds), *Clones and Clones: Facts and Fantasies about Human Cloning*, New York, W.W. Norton and Co., 54–66.

Dean, Mitchell (1994), *Critical and Effective Histories: Foucault's Methods and Historical Sociology*, London, Routledge.

Deery, June (1994), 'Ectopic and Utopic Reproduction: He, She and It', *Utopian Studies*, 5:2, 36–49.

Denny, Charlotte (1999), 'Cyber Utopia? Only the Usual Candidates Need Apply', *Guardian* (online), available at http://www.newsunlimited.co.uk/print/0,3838,388249300.html [accessed 12 July 1999].

Derrida, Jacques (1967), *Of Grammatology*, trans. Gayatri Chakravorty Spivak, Baltimore, The Johns Hopkins University Press.

—— (1978), *Writing and Difference*, trans. A. Bass, Chicago, University of Chicago Press.

Dery, Mark (1992), 'Cyberculture', *South Atlantic Quarterly*, 91:4, 501–23.

—— (1993), 'Flame Wars', *South Atlantic Quarterly*, 92:4, 559–68.

Dibbell, Julian (1998), *My Tiny Life: Crime and Passion in a Virtual World*, London, Fourth Estate.

Dick, Philip K. (1968), *Do Androids Dream of Electric Sheep?* New York, Del Rey.

Downey, Gary Lee, Joseph Dumit and Sarah Williams (1995), 'Cyborg Anthropology', *Cultural Anthropology*, 10:2, 264–9.

Dozois, Gardner and Stanley Schmidt (1998), *Roads Not Taken: Tales of Alternate History*, New York, Del Rey.

Dreyfus, Hubert L. and Paul Rabinow (1982), *Michel Foucault: Beyond Structuralism and Hermeneutics*, London, Harvester Wheatsheaf.

DuBois, Page (1991), *Centaurs and Amazons: Women and the Pre-History of the Great Chain of Being*, Ann Arbor, Michigan.

Dupré, John (1991), 'Reflections on Biology and Culture', in J.J. Sheehan and M. Sosna (eds), *The Boundaries of Humanity: Humans, Animals, Machines*, Berkeley, University of California Press, 125–31.

Ellul, Jacques (1965), *The Technological Society*, trans. J. Wilkinson, intro. R. Merton, London, Jonathan Cape.

Engels, F. [1842] (1958), *The Condition of the Working Class in England*, ed. and trans. W.O. Challoner and W.H. Henderson, Stanford, CA, University of California Press.

Eskridge, William N. and Edward Stein (1998), 'Queer Clones', in M.C. Nussbaum and C.R. Sunstein (eds), *Clones and Clones: Facts and Fantasies about Human Cloning*, New York, W.W. Norton and Co., 95–113.

Evans, Gavin (1998), 'Man Made', *Guardian*, London, 2 September, G2, 2–3.

Extropy Institute (1996), 'Extropians FAQ List: What Do "Transhuman" and "Posthuman" Mean?' (online), available at http://www.extropy.org/faq/transpost.html [accessed 19 March 1999].

Farley, Wendy (1990), *Tragic Vision and Divine Compassion*, Minneapolis, Fortress Press.

Farnell, Ross (1999), 'In Dialogue with "Posthuman" Bodies: Interview with Stelarc', *Body & Society*, 5:2–3, 129–47.

Featherstone, Mike and Roger Burrows (1995), 'Cultures of Technological Embodiment', *Body & Society*, 1:3–4, 1–19.

Feenberg, Andrew (1991), *Critical Theory of Technology*, New York, Oxford University Press.

Fine, Lawrence (1984), 'Kabbalistic Texts', in B.W. Holtz (ed.), *Back to the Sources: Reading the Classic Jewish Texts*, New York, Touchstone, 305–59.

Fitting, Peter (1991), 'The Lessons of Cyberpunk', in C. Penley and A. Ross (eds), *Technoculture*, Minneapolis, University of Minnesota Press, 295–315.

—— (1994), 'Beyond the Wasteland: A Feminist in Cyberspace', *Utopian Studies*, 5:2, 4–15.

Florescu, Radu (1976), *In Search of Frankenstein: Exploring the Myths behind Mary Shelley's Frankenstein*, Boston, New York Graphic Society.

Foerst, Anne (1996), 'Artificial Intelligence: Walking the Boundary', *Zygon: Journal of Religion and Science*, 31:4, 681–93.

—— (1997) ' "Cog", A Humanoid Robot and the Question of Imago Dei' (online), available at http://www.ai.mit.edu/people/annef/naturalism/naturalism.html [accessed 4 April 1999].

—— (1998) 'Embodied AI, Creation, and Cog', *Zygon: Journal of Religion and Science*, 33:3, 455–61.

Foster, Thomas (1999), 'Meat Puppets or Robopaths? Cyberpunk and the Question of Embodiment', in J. Wolmark (ed.), *Cybersexualities: A Reader on Feminist Theory, Cyborgs and Cyberspace*, Edinburgh, Edinburgh University Press, 208–29.

Foucault, Michel [1961] (1965), *Madness and Civilization*, trans. A. Howard, New York, Pantheon Press.

—— (1970), *The Order of Things*, London, Tavistock.

—— (1977), *Discipline and Punish: The Birth of the Prison*, trans. A. Sheridan, London, Penguin.

—— [1976] (1979), *The History of Sexuality Volume 1: An Introduction*, trans. R. Harley, London, Allen Lane.

—— (1980), *Power/Knowledge: Selected Interviews and Other Writings*, ed. C. Gordon, New York, Pantheon Press.

—— (1991a), 'What Is Enlightenment?' in P. Rabinow (ed.), *The Foucault Reader*, London, Penguin, 32–50.

—— [1971] (1991b), 'Nietzsche, Genealogy, History', in P. Rabinow (ed.), *The Foucault Reader*, London, Penguin, 76–100.

—— (1991c), 'What Is an Author?' in P. Rabinow (ed.), *The Foucault Reader*, 101–20.

—— (1991d) 'On the Genealogy of Ethics', in P. Rabinow (ed.), *The Foucault Reader*, 340–72.

—— (1999) [1979], 'Pastoral Power and Political Reason', in J. Carrette (ed.), *Religion and culture by Michel Foucault*, Manchester, Manchester University Press, 135–52.

<Frank@knarf.demon.co.uk> (1998), 'Frequently Asked Questions: What Is Cyberpunk?' *Alt.Cyberpunk* (online), available at http://www.knarf.demon.co.uk [accessed 11 September 1998].

Franklin, Sarah Brooks (1988), 'Life Story: The Gene as Fetish Object on TV', *Science As Culture*, 3, 92–100.

Frederickson, Eric (1998), 'Faith in the 24th Century', *Star Trek Monthly*, 41, July, 30–5.

Freud, Sigmund [1930] (1995), 'Civilization and its Discontents', in P. Gay (ed.), *The Freud Reader*, London, Vintage, 722–72.

Fuchs, Cynthia (1995), ' "Death is Irrelevant": Cyborgs, Reproduction and the Future of Male Hysteria', in C.H. Gray (ed.), *The Cyborg Handbook*, London, Routledge, 281–300.

Fulkerson, Mary McClintock and Susan Dunlap (1998), 'Michel Foucault', in G. Ward (ed.), *The Postmodern God*, Oxford, Blackwell, 116–23.

Galison, Peter (1994), 'The Ontology of the Enemy: Norbert Wiener and the Cybernetic Vision', *Critical Inquiry*, 21, 228–66.

Gaskell, Elizabeth [1848] (1970), *Mary Barton*, ed. S. Gill, Harmondsworth, Penguin.

Gibson, William (1984), *Neuromancer*, London, Victor Gollancz.

—— (1986), 'The Gernsback Continuum', in B. Sterling (ed.), *Mirrorshades: The Cyberpunk Anthology*, New York, Arbor House, 1–13.

—— [1981] (1997), 'Johnny Mnemonic', in R. Alexander (ed.), *Cyber-Killers*, London, Orion, 341–56.

Gibson, William and Bruce Sterling (1990), *The Difference Engine*, London, Vista.

Gilbert, Sandra M. and Susan Gubar (1979), 'Horror's Twin: Mary Shelley's Monstrous Eve', in *The Madwoman in the Attic: The Woman Writer and the Nineteenth-Century Imagination*, New Haven, Yale University Press, 213–47.

Gilbert, Walter (1992), 'A Vision of the Grail', in D. Kevles and L. Hood (eds), *The Code of Codes: Scientific and Social Issues in the Human Genome Project*, Cambridge, MA, Harvard University Press, 83–98.

Gimbutas, M. (1989), 'Women and Culture in Goddess-Oriented Old Europe', in C.P. Christ and J. Plaskow (eds), *Weaving the Visions: New Patterns in Feminist Spirituality*, San Francisco, Harper & Row, 63–71.

Glover, Jonathan (1984), *What Sort of People Should there Be?* Harmondsworth, Penguin.

Goldberg, Jonathan (1995), 'Recalling Totalities: The Mirrored Stages of Arnold Schwarzenegger', in C.H. Gray (ed.), *The Cyborg Handbook*, London, Routledge, 233–54.

Goldenberg, Naomi (1995), 'The Return of the Goddess: Psychoanalytic Reflections on the Shift from Theology to Thealogy', in U. King (ed.), *Religion and Gender*, Oxford, Blackwell, 145–64.

Gonzáles, Jennifer (1995), 'Envisioning Cyborg Bodies: Notes from Current Research', in C.H. Gray, (ed.), *The Cyborg Handbook*, London, Routledge, 267–79.

Goodall, Jane (1999), 'An Order of Pure Decision: Un-Natural Selection in the Work of Stelarc and Orlan', *Body & Society*, 5:2–3, 149–70.

Goodeve, Thyrza Nichols (2000), *How Like a Leaf: An Interview with Donna J. Haraway*, New York: Routledge.

Gore, Albert (1994), 'Remarks to the Superhighway Summit' (online), UCLA, 11 January, available at http://artcontext.com/cal/97/superhig.txt [accessed 2 December 2000].

Graham, Elaine (1999), 'Cyborgs or Goddesses? Becoming Divine in a Cyberfeminist Age', *Information, Communication & Society*, 2:4, 419–38.

Graham, Gordon (1999), *The Internet: A Philosophical Inquiry*, London, Routledge.

Graubard, M. (1967), 'The Frankenstein Syndrome: Man's Ambivalent Attitude to Knowledge and Power', *Perspectives in Biology and Medicine*, spring, 418–45.

Gray, C.H. (ed.) (1995a), *The Cyborg Handbook*, London, Routledge.

—— (1995b), 'An Interview with Manfred Clynes', in C.H. Gray (ed.), *The Cyborg Handbook*, London, Routledge, 43–53.

—— (1997), 'The Ethics and Politics of Cyborg Embodiment: Citizenship as Hypervalue', *Cultural Studies*, 1:2, 252–8.

Gray, C.H. and S. Mentor (1995), 'The Cyborg Body Politic: Version 1.2', in C.H. Gray (ed.), *The Cyborg Handbook*, London, Routledge, 453–67.

Gray, C.H., S. Mentor and H.J. Figueroa-Sarriera (1995), 'Cyborgology: Constructing the Knowledge of Cybernetic Organisms', in C.H. Gray, (ed.), *The Cyborg Handbook*, London, Routledge, 1–14.

Green, Nichola (1997), 'Beyond Being Digital: Representation and Virtual Corporeality', in D. Holmes (ed.), *Virtual Politics: Identity and Community in Cyberspace*, London, Sage, 59–78.

Grossberg, Lawrence (1996), 'Identity and Cultural Studies – Is That All There Is?' in S. Hall and P. du Gay (eds), *Questions of Cultural Identity*, London, Sage, 87–107.

Grosz, Elizabeth (1989), *Sexual Subversions: Three French Feminists*, Sydney, Allen & Unwin.

Gruber, Michael (1997a), 'In Search of the Electronic Brain', *Wired Magazine* (online), available at http://www.wired.com/collections/robots_ai/5.05_symbolic_ai.html [accessed 11 September 1998].

—— (1997b), 'Map the Genome, Hack the Genome', *Wired Magazine* (online), available at http://www.wired.com/collections/genetics/5.10_genome_hack.html [accessed 11 September 1998].

Guardian (2000), 'Gay Couple's Twins Win Right to Stay', *Guardian* (online), available at http://www.newsunlimited.co.uk/print/0,3858, 3954934,00.html [accessed 26 January 2000].

Haber, Honi Fern (1994), *Beyond Postmodern Politics: Lyotard, Rorty, Foucault*, London, Routledge.

Habermas, J. (1970), *Toward a Rational Society*, trans. J. Shapiro, Boston, Beacon Press.

Hacking, Ian (1986), 'The Archaeology of Foucault', in D. Hoy (ed.), *Foucault: A Critical Reader*, Oxford, Blackwell, 27–40.

Haining, Peter (ed.) (1994), *The Frankenstein Omnibus*, New Jersey, Chartwell.

Halberstam, Judith M. and Livingston, Ira (1995), 'Introduction: Posthuman Bodies', in J.M. Halberstam and I. Livingston (eds), *Posthuman Bodies*, Bloomington, Indiana University Press, 1–19.

Hall, Stuart (1996), 'Who Needs "Identity"'? in S. Hall and P. du Gay (eds), *Questions of Cultural Identity*, London, Sage, 1–17.

—— (1997a), Introduction in S. Hall (ed.), *Representation: Cultural Representation and Signifying Politics*, Buckingham, Open University, 1–11.

—— (1997b), 'The Work of Representation', in S. Hall (ed.), *Representation: Cultural Representation and Signifying Politics*, Buckingham, Open University, 15–63.

Hanley, R. (1997), *The Metaphysics of Star Trek*, San Francisco, Harper & Row.

Haraway, Donna (1987), 'Contested Bodies', in M. McNeil (ed.), *Gender and Expertise*, London, Free Association Books.

—— (1990), 'Investment Strategies for the Evolving Portfolio of Primate Females', in M. Jacobus, E. Fox Keller and S. Shuttleworth (eds), *Body/Politics: Women and the Discourse of Science*, London, Routledge, 139–62.

—— (1991a), 'A Cyborg Manifesto: Science, Technology, and Socialist-Feminism in the Late Twentieth Century', in *Simians, Cyborgs and Women: The Reinvention of Nature*, London, Free Association Books, 149–81.

—— (1991b), 'Situated Knowledges: The Science Question in Feminism and the Privilege of Partial Perspective', in *Simians, Cyborgs and Women: The Reinvention of Nature*, London, Free Association Books, 183–201.

—— (1991c), 'The Actors Are Cyborg, Nature Is Coyote, and the Geography Is Elsewhere: Postscript to "Cyborgs at Large"', in C. Penley and A. Ross (eds), *Technoculture*, Minneapolis, University of Minnesota Press, 21–6.

—— (1992a), 'The Promises of Monsters', in L. Grossberg, C. Nelson and P. Treichler (eds), *Cultural Studies*, New York, Routledge, 295–337.

—— (1992b), *Primate Visions: Gender, Race and Nature in the World of Modern Science*, 2nd edn, London, Verso.

—— (1992c), 'When Man™ Is on the Menu', in J. Crary and S. Kwinter (eds), *Incorporations*, London, Zone Books, 38–43.

—— (1993), 'Ecce Homo, Ain't (Ar'n't) I a Woman, and Inappropriate/d Others: The Human in a Post-Humanist Landscape', in J. Butler and J.W. Scott (eds), *Feminists Theorize the Political*, London, Routledge, 86–100.

—— (1994), 'A Game of Cat's Cradle: Science Studies, Feminist Theory, Cultural Studies', *Configurations*, 1, 59–71.

—— (1995), 'Cyborgs and Symbionts: Living Together in the New World Order', in C.H. Gray (ed.), *The Cyborg Handbook*, London, Routledge, xi–xx.

—— (1997), *Modest_Witness@Second_Millennium.FemaleMan©_Meets_OncoMouse™*, London, Routledge.

Haroontunian, H.D. (1988), 'Foucault, Genealogy, History: The Pursuit of Otherness', in J. Arac (ed.), *After Foucault: Humanistic Knowledge, Postmodern Challenges*, New Brunswick, NJ, Rutgers University Press, 110–37.

Harper, Mary Catherine (1995), 'Incurably Alien Other: A Case for Feminist Cyborg Writers', *Science-Fiction Studies*, 22, 399–420.

Harris, John (1998), *Clones, Genes, and Immortality*, Oxford, Oxford University Press.

Harrison, Taylor (1996a), 'Weaving the Cyborg Shroud: Mourning and Deferral in Star Trek: The Next Generation', in T. Harrison, S. Projansky, K.A. Ono and E.R. Helford (eds), *Enterprise Zones: Critical Positions on Star Trek*, Boulder, CO., Westview Press, 245–57.

—— (1996b), 'Interview with Henry Jenkins', in T. Harrison, S. Projansky, K.A. Ono and E.R. Helford (eds), *Enterprise Zones: Critical Positions on Star Trek*, Boulder, CO., Westview Press, 259–78.

Harrison, Taylor, Sarah Projansky, Kent A. Ono and Elyce Rae Helford (1996), Introduction in T. Harrison, S. Projansky, K.A. Ono and E.R. Helford (eds), *Enterprise Zones: Critical Positions on Star Trek*, Boulder, CO., Westview Press, 1–8.

Hart, Kevin (1997), 'Jacques Derrida', in G. Ward (ed.), *The Postmodern God*, Oxford, Blackwell, 159–67.

Hayles, N. Katherine (1993), 'The Life Cycle of Cyborgs: Writing the Posthuman', in M. Benjamin (ed.), *A Question of Identity: Women, Science and Literature*, New Brunswick, NJ, Rutgers University Press, 152–79.

—— (1996), 'Narratives of Artificial Life', in G. Robertson et al. (eds), *FutureNatural: Nature/Science/Culture*, London, Routledge, 146–64.

—— (1999), *How we Became Posthuman: Virtual Bodies in Cybernetics, Literature, and Informatics*, Chicago, University of Chicago Press.

Heelas, Paul (1998), *The New Age*, Oxford, Blackwell.

Heffernan, James A. (1997), 'Looking at the Monster: *Frankenstein* and Film', *Critical Inquiry*, 24, 133–58.

Heidegger, Martin (1962), *Being and Time*, trans. J. Macquarrie and E. Robinson, London, SCM Press.

—— [1966] (1976), 'Only a God Can Save us Now': An Interview with Martin Heidegger', *Philosophy Today*, 20:4, 267–84.

—— [1954] (1993a), 'The Question of Technology', in D.F. Krell (ed.), *Basic Writings*, London, Routledge, 307–42.

—— [1954] (1993b), 'Building Dwelling Thinking', in D.F. Krell (ed.), *Basic Writings*, London, Routledge, 343–63.

Heim, Michael (1993a), 'The Erotic Ontology of Cyberspace', in M. Benedikt (ed.), *Cyberspace: First Steps*, Cambridge, MA, MIT Press, 59–80.

—— (1993b), *The Metaphysics of Virtual Reality*, Oxford, Oxford University Press.

—— (1995), 'The Design of Virtual Reality', *Body & Society*, 1:3–4, 65–77.

Hewitt, Marsha A. (1993), 'Cyborgs, Drag Queens and Goddesses: Emancipatory-regressive Paths in Feminist Theory', *Method and Theory in the Study of Religion*, 5:2, 135–54.

—— (1994), 'The Redemptive Power of Memory: Walter Benjamin and Elizabeth Schüssler Fiorenza', *Journal of Feminist Studies in Religion*, 10, 73–89.

Hodges, Andrew (1992), *Alan Turing: The Enigma*, 2nd edn, London, Vintage.

Hogle, Jerrold E. (1995), 'Otherness in *Frankenstein*: The Confinement/Autonomy of Fabrication', in F. Botting (ed.), *Frankenstein: Contemporary Critical Essays*, London, Macmillan, 206–34.

Hollinger, Veronica (1999), 'Cybernetic Deconstructions: Cyberpunk and Postmodernism', in J. Wolmark (ed.), *Cybersexualities: A Reader on Feminist Theory, Cyborgs and Cyberspace*, Edinburgh, Edinburgh University Press, 174–90.

Holmes, David (1997a), Introduction in D. Holmes (ed.), *Virtual Politics: Identity and Community in Cyberspace*, London, Sage, 1–25.

—— (1997b), 'Virtual Identity: Communities of Broadcast, Communities of Interactivity', in D. Holmes (ed.), *Virtual Politics: Identity and Community in Cyberspace*, London, Sage, 26–45.

Holtz, Barry W. (1984), 'Midrash', in B.W. Holtz (ed.), *Back to the Sources: Reading the Classic Jewish Texts*, New York, Touchstone, 177–211.

Hood, Leroy (1992), 'Biology and Medicine in the Twenty-First Century', in D. Kevles and L. Hood (eds), *The Code of Codes: Scientific and Social Issues in the Human Genome Project*, Cambridge, MA, Harvard University Press, 136–63.

Hoy, David Couzens (1986), Introduction in D.C. Hoy (ed.), *The Foucault Reader*, Oxford, Blackwell, 1–25.

—— (1988), 'Foucault: Modern or Postmodern?' in J. Arac (ed.), *After Foucault: Humanistic Knowledge, Postmodern Challenges*, New Brunswick, NJ, Rutgers University Press, 12–41.

Hubbard, R. and E. Wald (1997), *Exploding the Gene Myth*, Boston, Beacon Press.

Huet, Marie-Hélène (1993), *Monstrous Imagination*, Cambridge, MA, Harvard University Press.

Human Genome Project (1999), 'FAQ: What Is the Human Genome Project?' (online), available at http://www.ornl.gov/TechResources/Human_Genome/faq/faqs1.html [accessed 3 February 1999].

Hurley, Kelly (1995), 'Reading Like an Alien: Posthuman Identity in Ridley Scott's *Alien* and David Cronenberg's *Rabid*', in J.M. Halberstam and I. Livingston (eds), *Posthuman Bodies*, Bloomington, Indiana University Press, 203–24.

Idel, Moshe (1988), 'The Golem in Jewish Magic and Mysticism', in E.D. Bilski (ed.), *Golem! Danger, Deliverance and Art*, New York, New York Jewish Museum, 15–35.

—— (1990), *Golem: Jewish Magical and Mystical Traditions on the Artificial Anthropoid*, Albany, NY, State University of New York Press.

Ihde, Don (1998), 'Bodies, Virtual Bodies and Technology', in D. Welton (ed.), *Body and Flesh: A Philosophical Reader*, Oxford, Blackwell, 349–57.

Ingraffia, Brian D. (1995), *Postmodern Theory and Biblical Theology: Vanquishing God's Shadow*, Cambridge, Cambridge University Press.

Irigaray, L. (1987), 'Divine Women', in *Sexes and Genealogies* ed. and trans. G. Gill, New York, Columbia University Press, 57–72.

—— (1993a), *This Sex which Is Not One*, trans. C. Porter, Ithaca, NY, Cornell University Press.

—— (1993b), 'Women: Equal or Different?' in *Je, tu, nous: Toward a Culture of Difference*, London, Routledge, 12–14.

—— [1979] (1998), 'Equal to Whom?' in G. Ward (ed.), *The Postmodern God*, Oxford, Blackwell, 198–214.

Jabès, E. (1993), *The Book of Margins*, trans. R. Waldrop, intro. Mark C. Taylor, Chicago, University of Chicago Press.

James, Philip, Hugh Pennington and Derek Burke (1999), 'GM Foods Served at Doom Temperature', *Times Higher Education Supplement*, 19 February, 26–7.

Jantzen, G. (1998), *Becoming Divine: Towards a Feminist Philosophy of Religion*, Manchester, Manchester University Press.

Jenkins, Stephen (ed.) (1981), *Fritz Lang: The Image and the Look*, London, British Film Institute.

Jensen, Paul M. (1969), *The Cinema of Fritz Lang*, New York, A.S. Barnes & Co.

—— (ed.) (1979), *Metropolis*, London, Faber & Faber. (Film transcript.)

Jewish Publication Society (1985), *Tanakh: A New Translation of the Holy Scriptures According to the Traditional Hebrew Text*, Jerusalem, Jewish Publication Society.

Jindra, Michael (1994), 'Star-Trek Fandom as a Religious Phenomenon', *Sociology of Religion*, 55, 27–51.

—— (1999), '"Star Trek to me Is a Way of Life": Fan Expressions of Star Trek Philosophy', in J.E. Porter and D.L. McLaren (eds), *Star Trek and Sacred Ground: Explorations of Star Trek, Religion and American Culture*, Albany, NY, State University of New York Press, 217–30.

Jolly, Richard (ed.) (1999), *Human Development Report 1999*, New York, Oxford University Press.

Jones, Steve (1993), *The Language of the Genes*, London, HarperCollins.

Jordanova, Ludmilla (1986a), Introduction in L.J. Jordanova (ed.), *Languages of Nature: Critical Essays on Science and Literature*, London, Free Association Books, 15–47.

—— (1986b), 'Naturalizing the Family: Literature and the Bio-Medical Sciences in the Eighteenth Century', in L.J. Jordanova (ed.), *Languages of Nature: Critical Essays on Science and Literature*, London, Free Association Books, 86–116.

—— (1989), *Sexual Visions: Images of Gender in Science and Medicine between the Eighteenth and Twentieth Centuries*, London, Harvester Wheatsheaf.

—— (1994), 'Melancholy Reflections: Constructing an Identity for Unveilers of Nature', in S. Bann (ed.), *Frankenstein, Creation and Monstrosity*, London, Reaktion Books, 60–76.

Kahn, Axel (1997), 'Clone Mammals . . . Clone Man?' *Nature* (online), available at http://www.nature.com/Nature2/serve?SID=25602728&CAT=NatGen&PG=sheep/sheep5.html [accessed 18 September 1998].

Kaku, Michio (1998), *Visions: How Science Will Revolutionize the 21st Century and Beyond*, Oxford, Oxford University Press.

Karnicky, Jeff (1996), 'Beyond Polemics: Posthuman Readings', *Socialist Review*, 26:1–2, 175–79.

Keller, Evelyn Fox (1991), 'Language and Ideology in Evolutionary Theory: Reading Cultural Norms into Natural Law', in J.J. Sheehan and M. Sosna (eds), *The Boundaries of Humanity: Humans, Animals, Machines*, Berkeley, University of California Press, 85–102.

243

References

—— (1992), 'Nature, Nurture, and the Human Genome Project', in D. Kevles and L. Hood (eds), *The Code of Codes: Scientific and Social Issues in the Human Genome Project*, Cambridge, MA, Harvard University Press, 281–99.

Kellner, Douglas (1989), *Jean Baudrillard: From Marxism to Postmodernism and Beyond*, Stanford, CA, Stanford University Press.

Kelly, Kevin (1994), *Out of Control? The New Biology of Machines*, London, Fourth Estate.

Kevles, Daniel (1992), 'Out of Eugenics: The Historical Politics of the Human Genome', in D. Kevles and L. Hood (eds), *The Code of Codes: Scientific and Social Issues in the Human Genome Project*, Cambridge, MA, Harvard University Press, 3–36.

Kieval, Hillel J. (1997), 'Pursuing the Golem of Prague: Jewish Culture and the Invention of a Tradition', *Modern Judaism*, 17, 1–23.

King, Helen (1995), 'Half-Human Creatures', in J. Cherry (ed.), *Mythical Beasts*, London, British Museum Press, 138–67.

Kirby, Vicki (1997), *Telling Flesh: The Substance of the Corporeal*, London, Routledge.

Klima, Ivan (1998), 'The Birth of the Robot', *Independent on Sunday*, London, 6 September, B2.

Kolcaba, Raymond (2000a), 'Loss of the World: A Philosophical Dialogue', *Ethics and Information Technology*, 2:1, 3–9.

—— (2000b), 'Angelic Machines: A Philosophical Dialogue', *Ethics and Information Technology*, 2:1, 11–17.

Korzeniowska, Victoria B. (1996), 'Engaging with Gender: Star Trek's Next Generation', *Journal of Gender Studies*, 5:1, 19–25.

Kracauer, Siegfried (1979), 'Industrialism and Totalitarianism' in P.M. Jensen (ed.), *Metropolis*, London, Faber & Faber, 15–17.

Krauss, Lawrence M. (1996), *The Physics of Star Trek*, London, HarperCollins.

Kristeva, Julia (1982), *Powers of Horror: An Essay in Abjection*, trans. L.S. Roudiez, New York, Columbia University Press.

—— (1991), *Strangers to Ourselves*, trans. L.S. Roudiez, New York, Columbia University Press.

Kroker, Arthur and Michael Weinstein (1994), 'The Hyper-Texted Body, or Nietzsche Gets a Modem', *CTheory* (online), available at http://www.ctheory.com/e-hyper-texted.html [accessed 14 July 1999].

Kumar, Krishan (1985), 'Religion and Utopia', *Centre for the Study of Religion and Society Pamphlet Library*, 8, Canterbury, University of Kent at Canterbury.

Kunzru, Hari (1997), 'You Are Cyborg', *Wired Magazine* (online), available at http://www.wired.com/wired/archive/5.02/ffharaway_pr.html [accessed 3 February 1999].

Kuryllo, Helen A. (1994), 'Cyborgs, Sorcery, and the Struggle for Utopia', *Utopian Studies*, 5:2, 50–5.

Kurzweil, Ray (1999a), *The Age of Spiritual Machines*, London, Orion.

—— (1999b), 'The Coming Merging of Mind and Machine', *Scientific American*, 10:3, 56–61.

Latour, Bruno (1993), *We Have Never Been Modern*, trans. C. Porter, Hemel Hempstead, Harvester Wheatsheaf.

Law, John (1991), 'Introduction: Monsters, Machines and Sociotechnical Relations', in J. Law (ed.), *A Sociology of Monsters: Essays on Power, Technology and Domination*, London, Routledge, 1–23.

Lelwica, Michelle M. (1998), 'From Superstition to Enlightenment in the Race for Pure Consciousness: Antireligious Currents in Popular and Academic Feminist Discourse', *Journal of Feminist Studies in Religion*, 14:2, 108–23.

Lesser, Wendy (1992), Introduction in Mary Shelley, *Frankenstein, or the Modern Prometheus*, London, Century, v–xviii.

Levitas, Ruth (1993), 'The Future of Thinking about the Future', in J. Bird et al. (eds), *Mapping the Futures: Local Cultures, Global Change*, London, Routledge, 257–66.

Lewontin, Richard (1993), *The Doctrine of DNA: Biology as Ideology*, Harmondsworth, Penguin.

Lieb, Michael (1998), *Children of Ezekiel: Aliens, UFOs, the Crisis of Race, and the Advent of End Time*, Durham, NC, Duke University Press.

Linafelt, Tod (1997), 'Margins of Lamentation: Or, the Unbearable Whiteness of Reading', in T.K. Beal and D.M. Gunn (eds), *Reading Bibles, Writing Bodies*, London, Routledge, 219–31.

'Locutus' (1996), 'Identity, Paranoia, and Technology on *The Next Generation*' (online), available at http://english~www.hss.cmu.edu/cyber.startrek.html [accessed 10 April 1999].

Lunn, P. (1993), 'Do Women need the GODDESS? Some Phenomenological and Sociological Reflections', *Feminist Theology*, 4, 17–38.

Lupton, Deborah (1995), 'The Embodied Computer/User', *Body & Society*, 1:3–4, 97–112.

Lykke, Nina (1996a), Introduction in N. Lykke and R. Braidotti (eds), *Between Monsters, Goddesses and Cyborgs: Feminist Confrontations with Science, Medicine and Cyberspace*, London, Zed Books, 1–10.

—— (1996b), 'Between Monsters, Goddesses and Cyborgs: Feminist Confrontations with Science', in N. Lykke and R. Braidotti (eds), *Between Monsters, Goddesses and Cyborgs: Feminist Confrontations with Science, Medicine and Cyberspace*, London, Zed Books, 13–29.

Lykke, Nina and Rosi Braidotti (1996), 'Postface', in N. Lykke and R. Braidotti (eds), *Between Monsters, Goddesses and Cyborgs: Feminist Confrontations with Science, Medicine and Cyberspace*, London, Zed Books, 242–9.

Lyon, David (1999), *Postmodernity*, 2nd edn, Buckingham, Open University.

Lyotard, Jean-François (1989), 'Can Thought Go on without a Body?' *Discourse* 11, 74–87.

McCarron, Kevin (1995), 'Corpses, Animals, Machines and Mannequins: The Body and Cyberpunk', *Body & Society*, 1:3–4, 261–73.

McCrickard, J.E. (1991), 'Born-Again Moon: Fundamentalism in Christianity and the Feminist Spirituality Movement', *Feminist Review*, 37, 59–67.

McGuire, William, (ed.) (1974), *The Freud/Jung Letters*, Princeton, Princeton University Press.

McLane, Maureen Noelle (1996), 'Literate Species: Populations, "Humanities", and "Frankenstein"', *English Literary History* 63, 959–88.

McNay, Lois (1992), *Foucault and Feminism*, Cambridge, Polity Press.

Madsen, C. (1994), 'A God of One's Own: Recent Work by and about Women in Religion', *Signs: Journal of Women in Culture and Society*, 19:2, 480–98.

Madsen, C. et al. (1989), 'Roundable Discussion: If God Is God She Is Not Nice', *Journal of Feminist Studies in Religion*, 5:1, 103–17.

Marsden, Jill (1996), 'Virtual Sexes and Feminist Futures: The Philosophy of "Cyberfeminism"', *Radical Philosophy*, 78, 6–16.

May, Stephen (1998), *Stardust and Ashes: Science Fiction in Christian Perspective*, London, SPCK.

Mazlish, Bruce (1993), *The Fourth Discontinuity: The Co-evolution of Humans and Machines*, New Haven, Yale University Press.

Mellor, Anne K. (1988), *Mary Shelley: Her Life, Her Fiction, Her Monsters*, London, Routledge.

—— (1995), 'A Feminist Critique of Science', in F. Botting (ed.), *Frankenstein: Contemporary Critical Essays*, London, Macmillan, 107–39.

Menser, Michael and Stanley Aronowitz (1996), 'On Cultural Studies, Science, and Technology', in S. Aronowitz, B. Martinsons and M. Menser (eds), *Technoscience and Cyberculture*, London, Routledge, 7–28.

Mettrie, Julien de la [1747] (1995), 'Man a Machine' in I. Kramnik (ed.), *The Portable Enlightenment Reader*, New York, Penguin, 202–9.

Meyrink, Gustav [1911] (1994), 'The Golem', in P. Haining (ed.), *The Frankenstein Omnibus*, New Jersey, Chartwell, 523–35.

Midgley, Mary (1992), *Science as Salvation: A Modern Myth and its Meaning*, London, Routledge.

Moers, Ellen (1977), *Literary Women: The Great Writers*, New York, Garden City Press.

Monsanto Corporation (1999), 'Frankenstein Food? Take Another Look' (online), available at http://www.searchmonsanto.com/monsanto-uk/frankenstein-foods.html [accessed 23 November 1999].

Moravec, Hans (1988), *Mind Children: The Future of Robot and Human Intelligence*, Cambridge, MA, Harvard University Press.

—— (1998), *Robot: Mere Machine to Transcendent Mind*, New York, Oxford University Press.

More, Max (1998), 'The Extropian Principles: A Transhumanist Declaration, Version 3.0', (online), available at http://extropy.org.extprn.html [accessed 19 March 1999].

Morton, N. (1989), 'The Goddess as Metaphoric Image', in C.P. Christ and J. Plaskow (eds), *Weaving the Visions: New Patterns in Feminist Spirituality*, San Francisco, Harper & Row, 111–18.

245

Mulkay, M. (1996), 'Frankenstein and the Debate over Embryo Research', *Science, Technology and Human Values*, 21, 157–76.

National Bioethics Advisory Commission (1998), 'The Science and Application of Cloning', in M.C. Nussbaum and C.R. Sunstein (eds), *Clones and Clones: Facts and Fantasies about Human Cloning*, New York, W.W. Norton and Co., 29–40.

Nelkin, Dorothy (1992), 'The Social Power of Genetic Information', in D. Kevles and L. Hood (eds), *The Code of Codes: Scientific and Social Issues in the Human Genome Project*, Cambridge, MA, Harvard University Press, 177–90.

Nellist, John G. and Elliot M. Gilbert (1999), *Understanding Modern Telecommunications and the Information Superhighway*, New York, Artech House.

Nelson, J.R. (1994), *On the New Frontiers of Genetics and Religion*, Grand Rapids. Eerdmans.

Neverow, Yara (1994), 'The Politics of Incorporation and Embodiment: *Woman on the Edge of Time* and *He, She and It* as Feminist Epistemologies of Resistance', *Utopian Studies*, 5:2, 16–35.

Newton, Polly, David Brown and Charles Clover (1999), 'Alarm over "Frankenstein Foods"', *Daily Telegraph*, 12 February, 1–2.

Nguyen, Dan Thu and Jon Alexander (1996), 'The Coming of Cyberspacetime and the End of the Polity', in R. Shields (ed.), *Cultures of Internet: Virtual Spaces, Real Histories, Living Bodies*, London, Sage, 99–124.

Noble, David F. (1999), *The Religion of Technology: The Divinity of Man and the Spirit of Invention*, 2nd edn, New York, Penguin.

Nuffield Council on Bioethics (1999), 'Genetically Modified Foods: The Ethical and Social Issues' (online), available at http://www.nuffieldfoundation.org [accessed 9 July 1999].

Nunes, Mark (1995), 'Baudrillard in Cyberspace: Internet, Virtuality, and Postmodernity', *Style*, 29, 314–27.

Nussbaum, M.C. and C.R. Sunstein (eds) (1998), 'Clones and Clones: Facts and Fantasies about Human Cloning', New York, W.W. Norton & Co.

Oates, Joyce Carol (1984), 'Frankenstein's Fallen Angel', *Critical Inquiry*, 10, 543–54.

O'Flinn, Paul (1995), 'Production and Reproduction', in F. Botting (ed.), *Frankenstein: Contemporary Critical Essays*, London, Macmillan, 21–47.

Okuda, Michael and Denise Okuda (1999), *The Star Trek Encyclopedia: A Reference Guide to the Future*, 3rd edn, New York, Pocket Books.

O'Leary, Stephen D. (1996), 'Cyberspace as Sacred Space: Communicating Religion on Computer Networks', *Journal of the American Academy of Religion*, 59:4, 781–808.

Oliver, Simon (1999), 'The Eucharist Before Nature and Culture', *Modern Theology*, 15:3, 331–53.

Paden, William (1994), *Religious Worlds*, 2nd edn, Boston, Beacon Press.

—— (2000), 'World', in W. Braun and R.T. McCutcheon (eds), *Guide to the Study of Religion*, London, Cassell, 334–47.

Paré, Ambroise [1573] (1982), *On Monsters and Marvels*, trans. and intro. Janis L. Pallister, Chicago, University of Chicago Press.

Pascoe, Eva and John L. Locke (2000), 'Can a Sense of Community Flourish in Cyberspace?' *Guardian* (online), available at
http://www.newsunlimited.co.uk/Print/0,3858,3972566,00.html [accessed 11 March 2000].

Pearson, Keith Ansell (1997), 'Life Becoming Body: On the "Meaning" of Post Human Evolution', *Cultural Studies*, 1:2, 219–40.

Penley, Constance (1997), *Nasa/Trek: Popular Science and Sex in America*, London, Verso.

Penley, Constance and Andrew Ross (1991a), Introduction in C. Penley and A. Ross (eds), *Technoculture*, Minneapolis, University of Minnesota Press, viii–xviii.

—— (1991b), 'Cyborgs at Large: Interview with Donna Haraway', in A. Penley and C. Ross (eds), *Technoculture*, Minneapolis: University of Minnesota Press, 1–20.

Peters, Ted (1997), *Playing God? Genetic Determinism and Human Freedom*, London, Routledge.

Piercy, Marge (1992), *Body of Glass*, London, Penguin.

—— (1994), 'Telling Stories about Stories', *Utopian Studies*, 5:2, 1–3.

—— (1995), *Eight Chambers of the Heart*, New York, Penguin.

Pitt, Joseph C. (1995), 'On the Philosophy of Technology, Past and Future', *Society for Philosophy & Technology*, 1:1/2, 1–8.

Plant, Sadie (1995), 'The Future Looms: Weaving Women and Cybernetics', Body & Society, 1:3–4, 45–64.

—— (1996a), 'On The Matrix: Cyberfeminist Simulations', in R. Shields (ed.), Cultures of Internet: Virtual Spaces, Real Histories, Living Bodies, London, Sage, 170–83.

—— (1996b), 'The Virtual Complexity of Culture', in G. Robertson, et al. (eds), FutureNatural: Nature/Science/Culture, London, Routledge, 203–17.

—— (1997), Zeros + Ones: Digital Women and the New Technoculture, London, Fourth Estate.

Porter, Jennifer E. and Darcee L. McLaren (eds) (1999), Star Trek and Sacred Ground: Explorations of Star Trek, Religion and American Culture, Albany, NY, State University of New York Press.

Porush, David (1989), 'Cybernetic Fiction and Postmodern Science', New Literary History, 20, 373–96.

—— (1998), 'Telepathy: Alphabetic Consciousness and the Age of Cyborg Illiteracy', in J. Broadhurst Dixon and E.J. Cassidy (eds), Virtual Futures: Cyberotics, Technology and Post-Human Pragmatism, London, Routledge, 45–64.

Poster, Mark (1996), 'Postmodern Virtualities', in G. Robertson, et al. (eds), FutureNatural: Nature/Science/Culture, London, Routledge, 183–202.

Postman, Neil (1993), Technopoly: The Surrender of Culture to Technology, New York, Vintage.

Pratchett, Terry (1996), Feet of Clay, London, Victor Gollancz.

Press association (1999), 'What Is the Future for Genetically Modified Crops?' (online), available at http://www.pa.press.net/news.backgrounder/gmfood/main/html [accessed 23 November 1999].

Pringle, David (ed.) (1997), The Ultimate Encyclopedia of Science Fiction, London, Carlton.

Prins, Bankje (1995), 'The Ethics of Hybrid Subjects: Feminist Constructivism According to Donna Haraway', Science, Technology and Human Values, 20:3, 352–67.

Puddefoot, John (1996), God and the Mind Machine: Computers, Artificial Intelligence and the Human Soul, London, SPCK.

Raphael, Melissa (1996a), Thealogy and Embodiment: The Post-Patriarchal Reconstruction of Female Sacrality, Sheffield, Sheffield Academic Press.

—— (1996b), 'Truth in Flux: Goddess Feminism as a Late Modern Religion', Religion, 26, 199–213.

Ray, Tom (1998), 'What Tierra Is' (online), available at http://www.hip.atr.co.jp/~ray/teirra/whatis.html [accessed 19 March 1999].

Reeves-Stevens, Judith and Garfield Reeves-Stevens (1997), Star Trek: The Next Generation: The Continuing Mission, New York, Pocket Books.

Regis, Ed (1990), Great Mambo Chicken and the Transhuman Condition: Science Slightly over the Edge, Reading, MA, Addison-Wesley.

—— (1994), 'Meet the Extropians', Wired Magazine (online), available at http://wired.com/wired/archive/2.10/extropians_pr.html [accessed 2 March 1999].

Rehmann-Sutter, Christoph (1996), 'Frankensteinian Knowledge?' Monist, 79, 264–79.

Reiss, M.J. and R. Straughan (1996), Improving Nature? The Science and Ethics of Genetic Engineering, Cambridge, UK, Cambridge University Press.

Rheingold, Howard (1991), Virtual Reality, London, Secker & Warburg.

Richards, Thomas (1997), Star Trek in Myth and Legend, London, Orion Media.

Roberts, Richard (1998), 'Time, Virtuality and the Goddess', Cultural Studies, 32:2/3, 270–87.

Roberts, Lynne D. and Malcolm R. Parks (1999), 'The Social Geography of Gender-Switching in Virtual Environments on the Internet', Information, Communication & Society, 2:4, 521–40.

Robins, Kevin (1995), 'Cyberspace and the World we Live in', Body & Society, 1:3–4, 135–55.

Robins, Ken and Les Levidow (1995), 'Socializing the Cyborg Self: The Gulf War and Beyond', in C.H. Gray (ed.), The Cyborg Handbook, London, Routledge, 119–25.

Robins, Kevin and Frank Webster (1988), 'Athens without Slaves . . . Or Slaves without Athens? The Neurosis of Technology', Science as Culture, 3, 7–53.

—— (1999), Times of the Technoculture, London, Routledge.

Rose, Hilary (1994), Love, Power and Knowledge: Towards a Feminist Transformation of the Sciences, Cambridge, Polity Press.

Rose, Mark (1956), Science Fiction: A Collection of Critical Essays, Engelwood Cliffs, NJ, Prentice-Hall.

Rose, Nikolas (1996), 'Identity, Genealogy, History', in S. Hall and P. du Gay (eds), Questions of Cultural Identity, London, Sage, 128–50.

Rosenthal, Pam (1991), 'Jacked in: Fordism, Cyberpunk, Marxism', *Socialist Review*, 21:1, 79–103.

Ross, Andrew (1991), *Strange Weather: Culture, Science and Technology in the Age of Limits*, London, Verso.

—— (1992), 'New Age Technoculture', in L. Grossberg, C. Nelson and P. Treichler (eds), *Cultural Studies*, New York, Routledge, 531–55.

—— (1996a), *Science Wars*, Durham, NC, Duke University Press.

—— (1996b), 'The Future is a Risky Business', in G. Robertson et al. (eds), *FutureNatural: Nature/ Science/Culture*, London, Routledge, 7–21.

Rouse, Joseph (1992), 'What Are Cultural Studies of Scientific Knowledge?' *Configurations*, 1:1, 57–94.

Rucker, Randy (1992), 'On the Edge of the Pacific', in R. Rucker, R.U. Sirius and Queen Mu (eds), *Mondo 2000: A User's Guide to the New Edge*, London, Thames & Hudson, 9–13.

Ruether, Rosemary R. (1992), *Sexism and God-Talk*, 2nd edn, Boston, Beacon Press.

Russ, Joanna (1995), *To Write Like a Woman: Essays in Feminism and Science Fiction*, Bloomington, Indiana University Press.

St. Clair, William (1989), *The Godwins and the Shelleys: The Biography of a Family*, London, Faber & Faber.

Sarup, Madan (1993), *An Introductory Guide to Post-Structuralism and Post-Modernism*, 2nd edn, Brighton, Harvester Wheatsheaf.

Schäfer, Peter (1995), 'The Magic of the Golem: The Early Development of the Golem Legend', *Journal of Jewish Studies*, 46:1–2, 249–61.

Scholem, Gershom (1965), 'The Idea of the Golem', in *On the Kabbalah and its Symbolism*, trans. Ralph Mannheim, New York, Schocken, 158–204.

—— (1971), 'Kabbalah', in *Encyclopedia Judaica*, Jerusalem, Keter Publishing, 490–653.

Schwab, Gabriele (1987), 'Cyborgs: Postmodern Phantasms of Body and Mind', *Discourse*, 9, 64–84.

Scully, Jackie Leach (1998), 'Women, Theology and the Human Genome Project', *Feminist Theology*, 17, 59–73.

Searle, John [1984] (1991), *Minds, Brains and Science*, London, Penguin.

Seltzer, Mark (1992), *Bodies and Machines*, London, Routledge.

Shakespeare, Tom (1998), 'Choices and Rights: Eugenics, Genetics and Disability Equality', *Disability and Society*, 13:5, 665–81.

Shaviro, Steven (1995), 'Two Lessons from Burroughs', in J.M. Halberstam and I. Livingston (eds), *Posthuman Bodies*, Bloomington, Indiana University Press, 38–54.

Sheehan, James J. and Morton Sosna (eds) (1991), *The Boundaries of Humanity: Humans, Animals, Machines*, Berkeley, University of California Press.

Shelley, Mary [1818] (1998), *Frankenstein: Or the Modern Prometheus*, ed. and intro. by Marilyn Butler, Oxford, Oxford University Press.

Sherwin, Byron L. (1985), *The Golem Legend: Origins and Implications*, Lanham: MD., University Press of America.

—— (1995), 'The Golem, Zevi Ashkenazi, and Reproductive Biotechnology', *Judaism*, 44, 314–22.

Shields, Rob, (1996), 'Virtual Spaces, Real Histories and Living Bodies', in R. Shields (ed.), *Cultures of Internet: Virtual Spaces, Real Histories, Living Bodies*, London, Sage, 1–10.

Shildrick, Margrit (1996), 'Posthumanism and the Monstrous Body', *Body & Society*, 2:1, 1–15.

—— (1997), *Leaky Bodies and Boundaries: Feminism, Postmodernism and (Bio)Ethics*, London, Routledge.

Simms, Andrew (ed.) (1999), *Selling Suicide: Farming, False Promises and Genetic Engineering in Developing Countries*, London, Christian Aid.

Singer, Isaac Bashevis (1988), Foreword in E.D. Bilski (ed.), *Golem! Danger, Deliverance and Art*, New York, New York Jewish Museum, 1–9.

Sirius, R.U. (1992), 'A User's Guide', in R. Rucker, R.U. Sirius and Queen Mu (eds), *Mondo 2000: A User's Guide to the New Edge*, London, Thames & Hudson, 14–20.

—— (1995), 'Mondo 2000 vs. Wired: Subjective Reflexions on the Righting of the New Edge', *21C* (online), Victoria, Australia, available at
http://www.scrappi.com/deceit/nrlydeep/mndvswir.html [accessed 11 September 1998].

Smith, Neil (1996), 'The Production of Nature', in G. Robertson et al. (eds), *FutureNatural: Nature/ Science/Culture*, London, Routledge, 35–54.

Sobchack, Vivian (1993), 'New Age Mutant Ninja Hackers: Reading *Mondo 2000*', *South Atlantic Quarterly*, 92:4, 569–84.

Sofia, Zoë (1992), 'Virtual Corporeality: A Feminist View', *Australian Feminist Studies*, 15, 11–24.

—— (1993), *Whose Second Self? Gender and (Ir)rationality in Computer Culture*, Geelong, Deakin University Press.

—— (1995), 'Of Spanners and Cyborgs: "Dehomogenizing" Feminist Thinking on Technology', in B. Caine and R. Pringle (eds), *Transitions: New Australian Feminisms*, St. Leonards, Allen & Unwin, 1995, 147–63.

—— (1997), 'Posthuman or Para-ego? Interactive Models of Human-Technology Relations', Conference Proceedings, 30 November 1997 (online), available at http://www.imago.com.au/WOV/papers/posthum.htm [accessed 3 March 1999].

Soper, Kate (1995), *What Is Nature? Culture, Politics and the Non-Human*, London, Routledge.

—— (1996), 'Nature/"Nature"', in G. Robertson et al. (eds), *FutureNatural: Nature/Science/Culture*, London, Routledge, 22–34.

—— (1999), 'Of OncoMice and FemaleMen: Donna Haraway on Cyborg Ontology', *Women: A Cultural Review*, 10:2, 167–72.

Sourbut, Elizabeth (1996), 'Gynogenesis: A Lesbian Appropriation of Reproductive Technologies', in N. Lykke and R. Braidotti (eds), *Between Monsters, Goddesses and Cyborgs: Feminist Confrontations with Science, Medicine and Cyberspace*, London, Zed Books, 227–41.

Spender, Dale (1995), *Nattering on the Net: Women, Power and Cyberspace*, Melbourne, Spinifex.

Springer, Claudia (1993), 'Sex, Memories, and Angry Women', *South Atlantic Quarterly*, 92:4, 713–33.

—— (1996), *Electronic Eros: Bodies and Desire in the Post-Industrial Age*, London, The Athlone Press.

Spufford, Francis (1996), 'The Difference Engine and *The Difference Engine*', in F. Spufford and J. Uglow (eds), *Cultural Babbage: Technology, Time and Invention*, London, Faber & Faber, 266–90.

Squier, Susan M. (1995), 'Reproducing the Posthuman Body: Ectogenetic Fetus, Surrogate Mother, Pregnant Man', in J.M. Halberstam and I. Livingston (eds), *Posthuman Bodies*, Bloomington, Indiana University Press, 113–32.

Steintrager, James A. (1997), 'Perfectly Inhuman: Moral Monstrosity in Eighteenth-Century Discourse', *Eighteenth-Century Life*, 21:2, 114–27.

Stelarc (1997), 'From Psycho to Cyber Strategies: Prosthetics, Robotics and Remote Existence', *Cultural Studies*, 1:2, 241–9.

—— (1998), 'From Psycho-Body to Cyber-Systems: Images as Post-Human Entities', in J. Broadhurst Dixon and E.J. Cassidy (eds), *Virtual Futures: Cyberotics, Technology and Post-Human Pragmatism*, London, Routledge, 116–23.

—— (1999), 'Parasite Visions: Alternate, Intimate and Involuntary Experiences', *Body & Society*, 5:2–3, 117–27.

Stenger, Nicole (1993), 'Mind Is a Leaking Rainbow', in M. Benedikt (ed.), *Cyberspace: First Steps*, Cambridge, MA, MIT Press, 49–58.

Stockton, K.B. (1994), *God between their Lips: Desire Between Women in Irigaray, Brontë and Eliot*, Stanford, CA, Stanford University Press.

Stone, Allucquere Rosanne (1993), 'Will the Real Body Please Stand up? Boundary Stories about Virtual Cultures', in M. Benedikt (ed.), *Cyberspace: First Steps*, Cambridge, MA, MIT Press, 81–118.

Starthern, Marilyn (1996), 'Enabling Identity? Biology, Choice and the New Reproductive Technologies', in S. Hall and P. du Gay (eds), *Questions of Cultural Identity*, London, Sage, 37–52.

—— (1998), 'Surrogates and Substitutes: New Practices for Old?' in J. Good and I. Velody (eds), *The Politics of Postmodernity*, Cambridge, UK, Cambridge University Press, 182–209.

Sutherland, John (1996), 'How Does Victor Make his Monsters?' in *Was Heathcliff a Murderer?* Oxford, Oxford University Press, 24–34.

Tattersall, Ian (1998), *Becoming Human: Evolution and Human Uniqueness*, Oxford, Oxford University Press.

Terranova, Tiziana (1996), 'Posthuman Unbounded: Artificial Evolution and High-tech Subcultures', in G. Robertson et al. (eds), *FutureNatural: Nature/Science/Culture*, London, Routledge, 165–80.

Todorov, Tzvetan (1975), *The Fantastic: A Structural Approach to a Literary Genre*, trans. R. Howard, Ithaca, N.Y., Cornell University Press.

Tomas, David (1993), 'Old Rituals for New Space: *Rites de Passage* and William Gibson's Cultural Model of Cyberspace', in M. Benedikt (ed.), *Cyberspace: First Steps*, Cambridge, MA, MIT Press, 31–41.

References

—— (1995), 'Feedback and Cybernetics: Reimaging the Body in the Age of the Cyborg', *Body & Society*, 1:3–4, 21–43.

Toumey, C. (1992), 'The Moral Character of Mad Scientists: A Cultural Critique of Science', *Science, Technology and Human Values*, 17, 411–37.

Trefil, James (1997), *Are we Unique?* New York, James Wiley & Sons.

Turkle, Sherry (1984), *The Second Self: Computers and the Human Spirit*, New York, Simon & Schuster.

—— (1991), 'Romantic Reactions: Paradoxical Responses to the Computer Presence', in J.J. Sheehan and M. Sosna (eds), *The Boundaries of Humanity: Humans, Animals, Machines*, Berkeley, University of California Press, 224–52.

—— (1995), *Life on the Screen: Identity in the Age of the Internet*, New York, Simon & Schuster.

Turney, Jon (1998), *Frankenstein's Footsteps: Science, Genetics and Popular Culture*, New Haven, Yale University Press.

'Vadi Mecum' (1998), *Myths and Monsters: Unravelling the Truth*, Bristol, Vadi Mecum Guides.

Vasseleu, Carolyn (1994), 'Virtual Bodies/Virtual Worlds', *Australian Feminist Studies*, 19, 155–69.

van den Brock, R. (ed.) (1997), *Gnosis and Hermeticism from Antiquity to Modern Times*, New York, State University of New York Press.

Vines, G. (1997), 'In and out of This World', *Times Higher Education Supplement*, 25 April, 17.

Virilio, Paul (1989), 'The Last Vehicle', in *Looking Back on the End of the World*, ed. D. Kamper and C. Wulf, New York, Sémiotext(e), 106–19.

Ward, Graham (1999), 'The Secular City and the Christian Corpus', *Cultural Studies*, 3:2, 140–63.

Ward, Mark (1999), *Virtual Organisms*, London, Macmillan.

Warwick, Kevin (1998), *In the Mind of the Machine: The Breakthrough in Artificial Intelligence*, London, Arrow.

Watson, James (1992), 'A Personal View of the Project', in D. Kevles and L. Hood (eds), *The Code of Codes: Scientific and Social Issues in the Human Genome Project*, Cambridge, MA, Harvard University Press, 164–73.

Wertheim, Margaret (1997), *Pythagoras' Trousers: God, Physics and the Gender Wars*, London, Fourth Estate.

—— (1999), *The Pearly Gates of Cyberspace*, London, Virago.

Wiener, Norbert (1948), *Cybernetics: Communication and Control in Animal and Machine*, Cambridge, MA, MIT Press.

—— (1964), *God and Golem, Inc. A Comment on Certain Points where Cybernetics Impinges on Religion*, Cambridge, MA, MIT Press.

—— [1950] (1989), *The Human Use of Human Beings: Cybernetics and Society*, ed. S.J. Heinz, London, Free Association Books.

Wilcox, Rhonda V. (1996), 'Dating Data: Miscegenation in *Star Trek: The Next Generation*', in T. Harrison, S. Projansky, K.A. Ono and E.R. Helford (eds), *Enterprise Zones: Critical Positions on Star Trek*, Boulder, CO., Westview Press, 69–92.

Williams, Raymond (1976), *Keywords: A Vocabulary of Culture and Society*, London, Fontana.

—— (1986), Foreword in L.J. Jordanova (ed.), *Languages of Nature: Critical Essays on Science and Literature*, London, Free Association Books, 10–14.

Williams, Rowan (2000), 'The Nature of a Sacrament', in *On Christian Theology*, Oxford, Blackwell, 197–208.

Wilmut, Ian, et al. (1997), 'Viable Offspring Derived from Fetal and Adult Mammalian Cells', *Nature*, 385, 27 February, 810–13.

Wilmut, Ian, Keith Campbell and Colin Tudge (2000), *The Second Creation: The Age of Biological Control*, London, Headline.

Wilson, Robert R. (1995), 'Cyber(body)parts: Prosthetic Consciousness', *Body & Society*, 1:3–4, 239–59.

Winograd, Terry (1991), 'Thinking Machines: Can there be? Are we?' in J.J. Sheehan and M. Sosna (eds), *The Boundaries of Humanity: Humans, Animals, Machines*, Berkeley, University of California Press, 198–224.

Witwer, Julia (1995), 'The Best of both Worlds: On *Star Trek's* Borg', in G. Brahm, Jr. and Mark Driscoll (eds), *Prosthetic Territories: Politics and Hypertechnologies*, Boulder, CO., Westview Press, 270–9.

Wolin, Richard (1994), 'Walter Benjamin: An Aesthetic of Redemption', Berkeley, University of California Press.

Wood, Virginia B. (1999), 'It's not Nice to Fool Mother Nature', *Austin Chronicle* (online), available at http://www.weeklywire.com/ww/08–09–99/austin_food_featuref.html [accessed 23 November 1999].

Yates, Frances (1979), *Giordano Bruno and the Hermetic Tradition*, Chicago, University of Chicago Press.

Ziguras, Christopher (1997), 'The Technologization of the Sacred: Virtual Reality and the New Age', in D. Holmes (ed.), *Virtual Politics: Identity and Community in Cyberspace*, London, Sage, 197–211.

Zimmerman, Michael E. (1990), *Heidegger's Confrontation with Modernity*, Bloomington, Indiana University Press.

Zorpette, Glenn (1999), 'Muscular Again', *Scientific American*, 10:3, 27–31.

Zurbrugg, Nicholas (1999), 'Marinetti, Chopin, Stelarc and the Auratic Intensities of the Postmodern Techno-Body', *Body & Society*, 5:2–3, 93–115.

Note: 'n.' after a page reference indicates a note number on that page.